Holt Mathematics

Chapter 10 Resource Book

HOLT, RINEHART AND WINSTON
A Harcourt Education Company
Orlando • Austin • New York • San Diego • London

Printed in the United States of America

ISBN 0-03-078227-9

4 5 170 09 08 07

CONTENTS

Blackline Masters

Parent Letter 1
Lesson 10-1 Practice A, B, C 3
Lesson 10-1 Reteach 6
Lesson 10-1 Challenge 8
Lesson 10-1 Problem Solving 9
Lesson 10-1 Reading Strategies 10
Lesson 10-1 Puzzles, Twisters & Teasers 11
Lesson 10-2 Practice A, B, C 12
Lesson 10-2 Reteach 15
Lesson 10-2 Challenge 16
Lesson 10-2 Problem Solving 17
Lesson 10-2 Reading Strategies 18
Lesson 10-2 Puzzles, Twisters & Teasers 19
Lesson 10-3 Practice A, B, C 20
Lesson 10-3 Reteach 23
Lesson 10-3 Challenge 24
Lesson 10-3 Problem Solving 25
Lesson 10-3 Reading Strategies 26
Lesson 10-3 Puzzles, Twisters & Teasers 27
Lesson 10-4 Practice A, B, C 28
Lesson 10-4 Reteach 31
Lesson 10-4 Challenge 32
Lesson 10-4 Problem Solving 33
Lesson 10-4 Reading Strategies 34
Lesson 10-4 Puzzles, Twisters, & Teasers 35
Lesson 10-5 Practice A, B, C 36
Lesson 10-5 Reteach 39
Lesson 10-5 Challenge 41
Lesson 10-5 Problem Solving 42
Lesson 10-5 Reading Strategies 43
Lesson 10-5 Puzzles, Twisters & Teasers 44
Lesson 10-6 Practice A, B, C 45
Lesson 10-6 Reteach 48
Lesson 10-6 Challenge 49
Lesson 10-6 Problem Solving 50
Lesson 10-6 Reading Strategies 51
Lesson 10-6 Puzzles, Twisters & Teasers 52
Lesson 10-7 Practice A, B, C 53
Lesson 10-7 Reteach 56
Lesson 10-7 Challenge 57
Lesson 10-7 Problem Solving 58
Lesson 10-7 Puzzles, Twisters & Teasers 60
Lesson 10-8 Practice A, B, C 61
Lesson 10-8 Reteach 64
Lesson 10-8 Challenge 65
Lesson 10-8 Problem Solving 66
Lesson 10-8 Reading Strategies 67
Lesson 10-8 Puzzles, Twisters & Teasers 68
Lesson 10-9 Practice A, B, C 69
Lesson 10-9 Reteach 72
Lesson 10-9 Challenge 73
Lesson 10-9 Problem Solving 74
Lesson 10-9 Reading Strategies 75
Lesson 10-9 Puzzles, Twisters & Teasers 76
Answers to Blackline Masters 77

Date ___________

Dear Family,

In this chapter, your child will learn about measuring the area and volume of various plane and solid geometric figures including the circle, rectangle, triangle, prism, pyramid, and cone. Whether you are designing the layout of a room, mapping a plot of land, or working as a physical scientist, understanding area and volume is integral to the work.

The **area** of a figure is the amount of surface the figure covers. Your child will learn how to estimate the area of irregular figures and how to find the area of rectangles, triangles, parallelograms, and trapezoids. Area is measured in square units.

Find the area of the shaded rectangle.

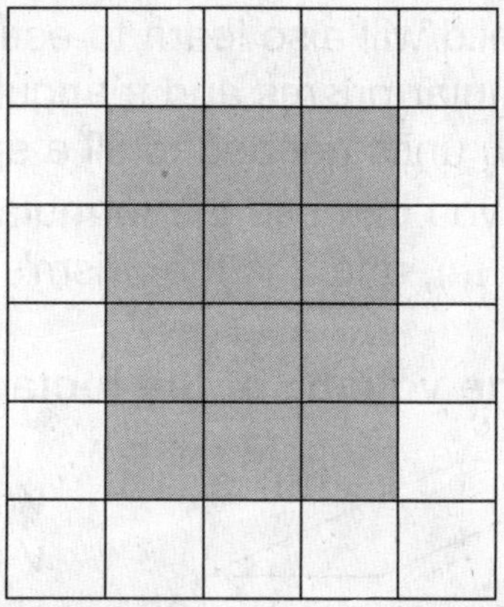

There are 4 rows of 3 shaded squares.

There are 12 shaded squares in all.

To find the area of a rectangle, multiply the length by the width.

$A = \ell w$

$A = 4 \cdot 3 = 12$

The area of the shaded rectangle is 12 square units.

Find the area of the triangle.

To find the area of a triangle, find one-half of the product of its base and its height.

$A = \frac{1}{2}bh$ *Write the formula.*

$A = \frac{1}{2}(12 \cdot 8)$ *Substitute 12 for b. Substitute 8 for h.*

$A = \frac{1}{2}(96)$ Multiply.

$A = 48$

The area is 48 cm^2

Your child will learn to find the area of a circle. To find the area of any circle, use the formula $A = \pi r^2$.

Find the area of the circle. Use $\frac{22}{7}$ for *pi*.

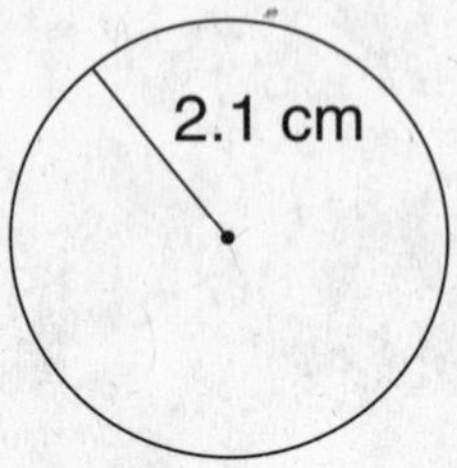

$A = \pi r^2$	*Write the formula for the area.*
$A \approx \frac{22}{7}(2.1)^2$	*Replace* π with $\frac{22}{7}$ *and r with 2.1.*
$A \approx \frac{22}{7} \cdot 4.41$	*Use the order of operations.*
$A \approx \frac{97.02}{7}$	*Multiply.*
$A \approx 13.86 \text{ cm}^2$	*Divide.*

Your child will also learn to estimate and find the volume of rectangular prisms and triangular prisms. **Volume** is the number of cubic units needed to fill a space. To find the volume of any prism, you can use the formula $V = Bh$, where B is the area of the base (ℓw), and h is the prism's height.

Find the volume of the rectangular prism.

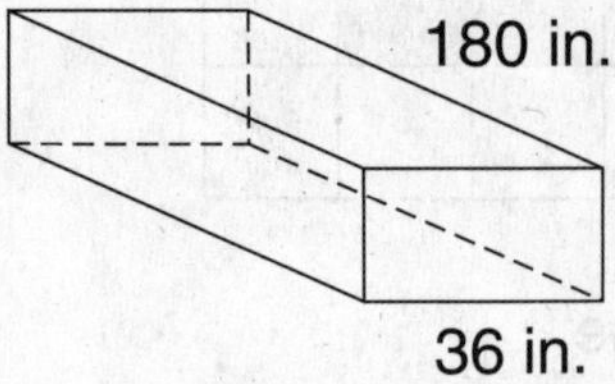

$V = \ell wh$	*Write the formula.*
$V = 180 \bullet 20 \bullet 36$	$\ell = 180$; $w = 20$; $h = 36$
$V = 129{,}600 \text{ in}^3$	*Multiply.*

For additional resources, visit go.hrw.com and enter the keyword MR7 Parent.

Name ______________________ Date ____________ Class ____________

LESSON 10-1

Practice A
Estimating and Finding Area

Estimate the area of each figure.

1.

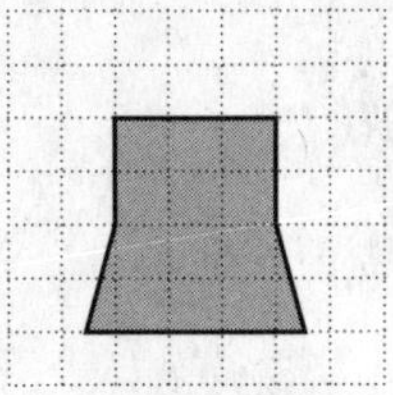

= 1 ft^2

____

2.

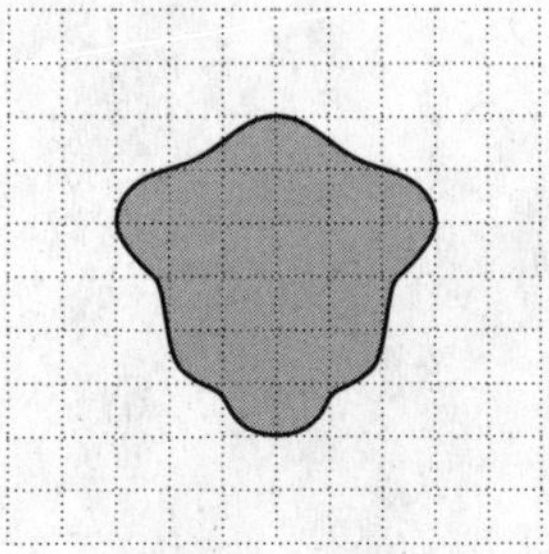

= 1 in^2

____

3.

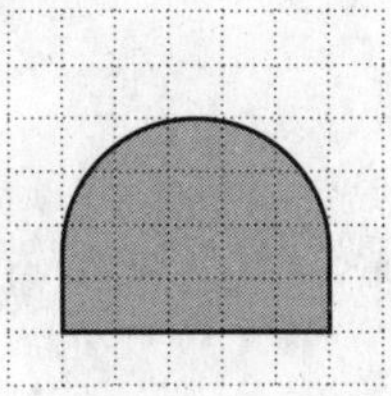

= 1 m^2

____

4.

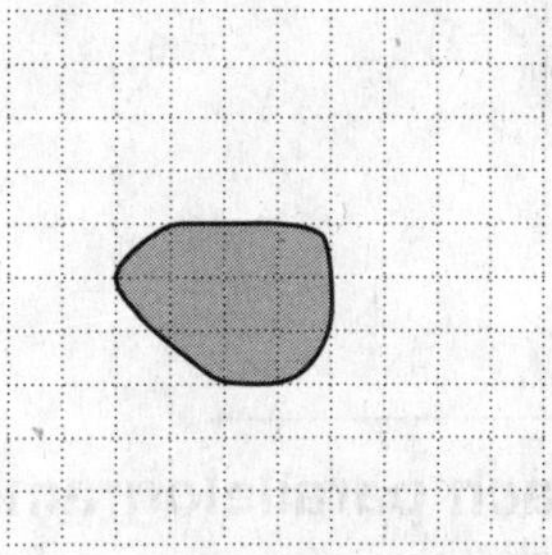

= 1 yd^2

____

Find the area of each rectangle.

5.

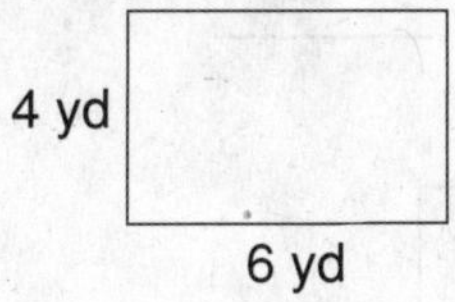

____

6.

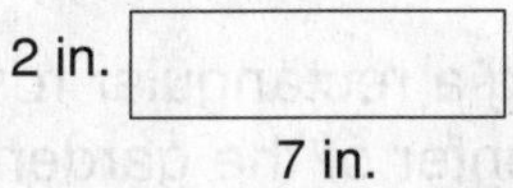

____

7.

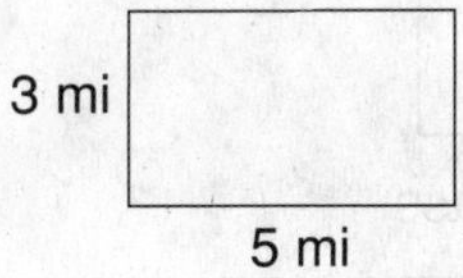

____

8.

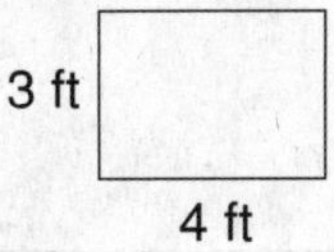

____

9. A square room has sides that each measure 5 feet. How many square feet of carpet is needed to cover the room's entire floor?

____

10. A rectangular coffee table is 2 feet wide and 4 feet long. How many square feet of glass is needed to cover the entire table top?

____

Name ______________________ Date ____________ Class ____________

LESSON **10-1**

Practice B

Estimating and Finding Area

Estimate the area of each figure.

1.

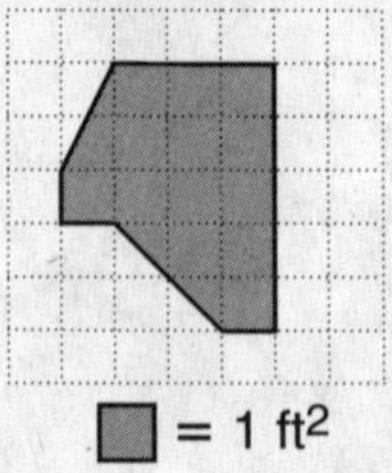

____

2.

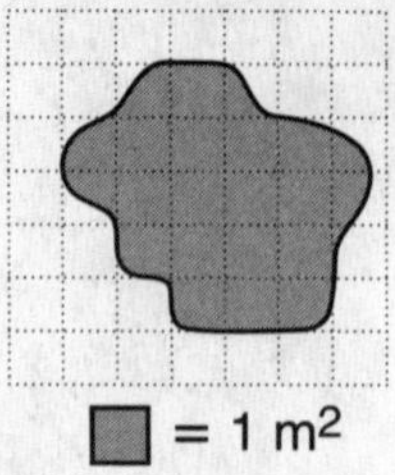

____

Find the area of each rectangle.

3.

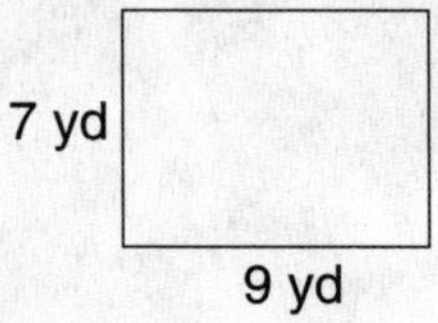

____

4.

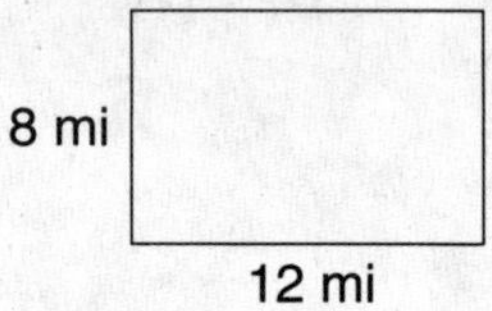

____

Find the area of each parallelogram.

5.

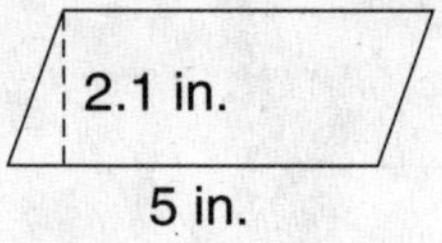

____

6.

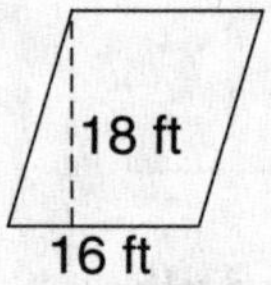

____

7. Mariah is planting a rectangular rose garden. In the center of the garden, she puts a smaller rectangular patch of grass. The grass is 2 ft by 3 ft. What is the area of the rose garden?

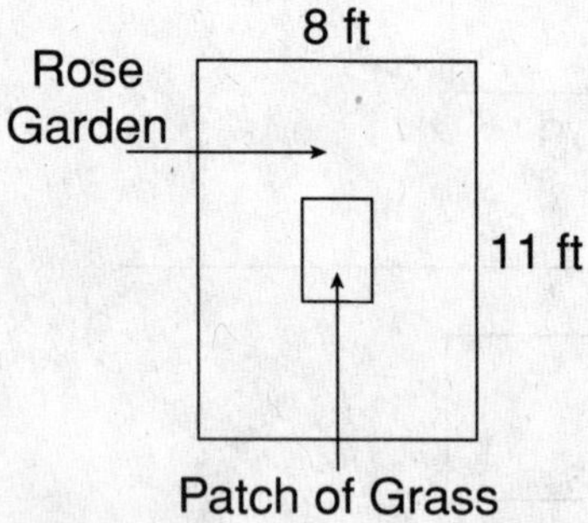

____

8. A section of a stained-glass window is shaped like a parallelogram. Its base is 6.5 inches, and its height is 4 inches. How much glass is needed to cover the section completely?

____

9. Your rectangular yard is 10 feet wide and 26 feet long. How many square feet of grass do you need to plant if you want to cover the entire yard?

____

Name ______________________ Date __________ Class __________

LESSON 10-1

Practice C
Estimating and Finding Area

Estimate the area of each figure.

1.

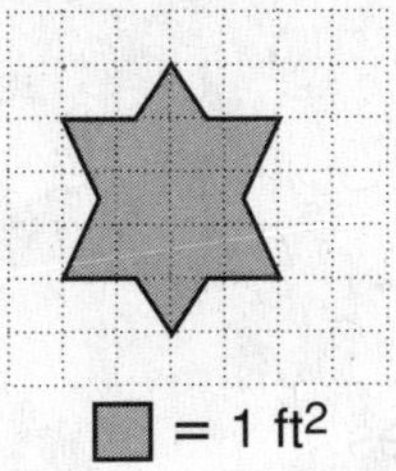

____

2.

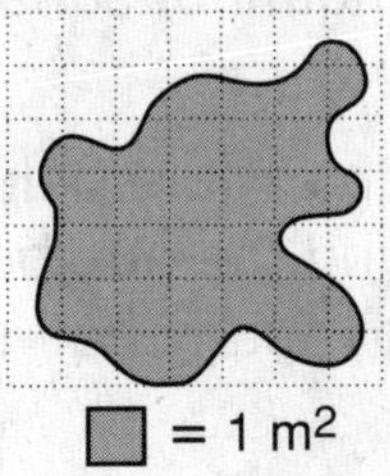

____

Find the area of each rectangle.

3.

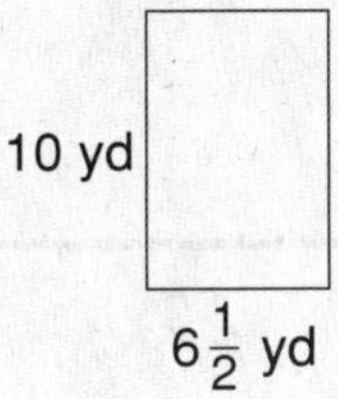

____

4.

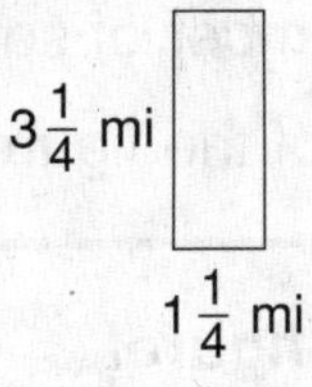

____

Find the area of each parallelogram.

5.

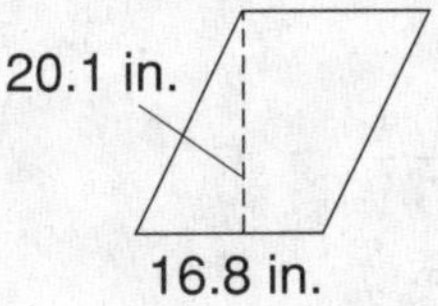

____

6.

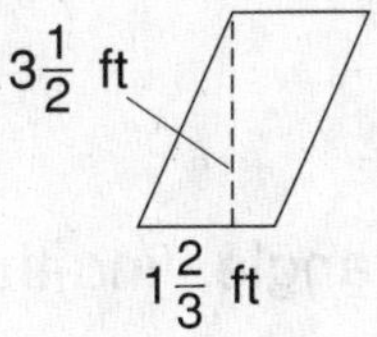

____

7. Denzel is building a square deck. On one side of the deck, he puts a smaller square hot tub. The hot tub is 5 ft long. What is the area of the deck?

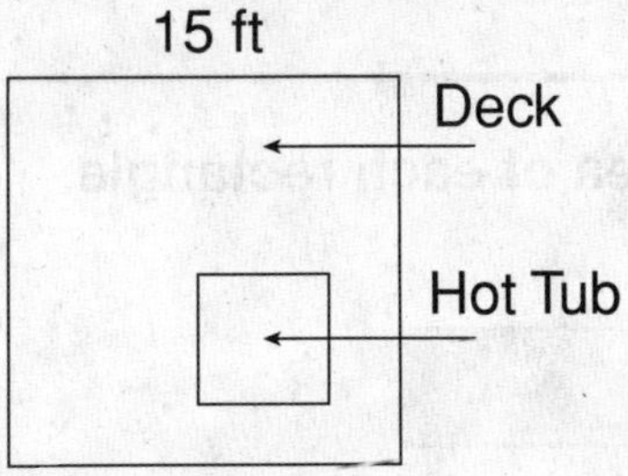

____

8. A kite is in the shape of a parallelogram with a 4.5 cm base and a height of 7.9 cm. What is the area of the kite?

____

Name ______________________ Date ____________ Class ____________

LESSON 10-1

Reteach

Estimating and Finding Area

You can use grid paper to estimate and find area of figures. The area of a figure is the amount of surface it covers. Area is measured in square units.

To estimate the area of the figure below, first find the number of fully shaded squares. Next, find the number of nearly full squares. Then, find the number of half or nearly half-shaded squares.

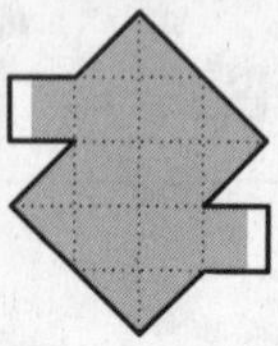

There are 6 fully shaded squares.
There are 2 nearly full squares.
There are 8 half-shaded squares.
$(8 \cdot \frac{1}{2} = 4)$

Find the estimated total number of squares shaded.

$6 + 2 + 4 = 12$. The area of the figure is about 12 square units.

Estimate the area of each figure.

1.

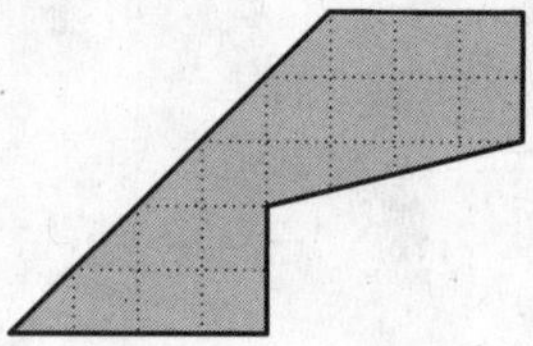

____

2.

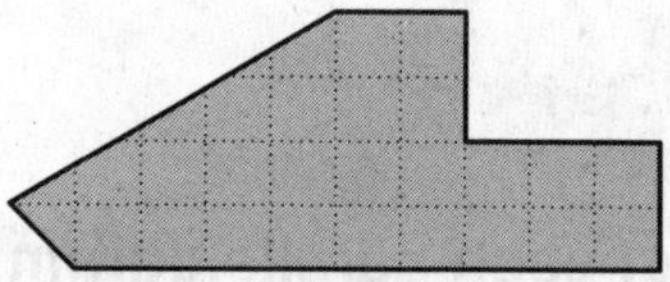

____

To find the area of a rectangle, find the total number of square units.

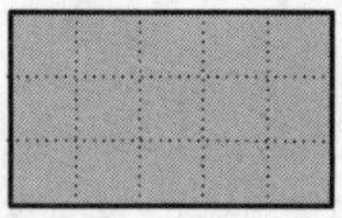

There are 3 rows of 5 squares.
$3 \cdot 5 = 15$
So, the area of the rectangle is 15 square units.

Find the area of each rectangle.

3.

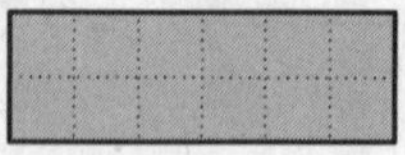

____

4.

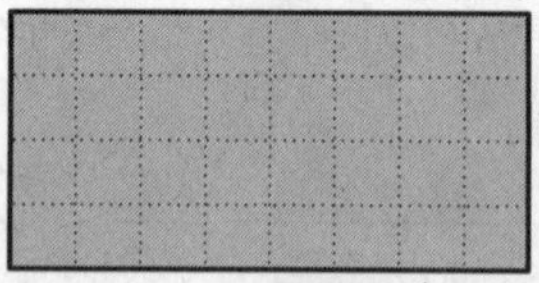

____

Name ______________________ Date ____________ Class ____________

LESSON **10-1**

Reteach

Estimating and Finding Area (continued)

To find the area of a parallelogram, first turn your parallelogram into a rectangle.

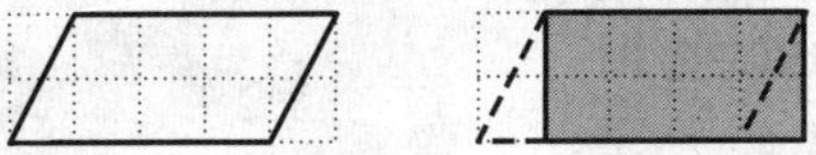

Then find the area of the rectangle.

Because the area of a rectangle is $A = \ell w$, the area of a parallelogram is $A = bh$.

$4 \cdot 2 = 8$

So, the area of the parallelogram is 8 square units.

Find the area of each parallelogram.

5.

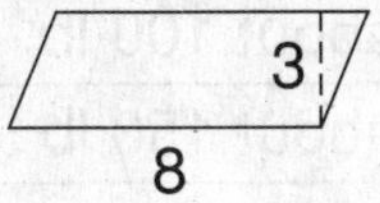

____

6.

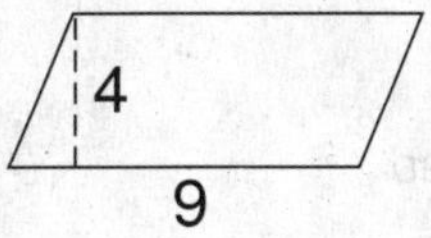

____

Sometimes you need to subtract a smaller area from a larger area.

A rectangular yard is made up of a rectangular flower garden surrounding a rectangular vegetable garden. The yard is 9 ft by 7 ft. The vegetable garden is 3 ft by 2 ft. What is the area of the flower garden?

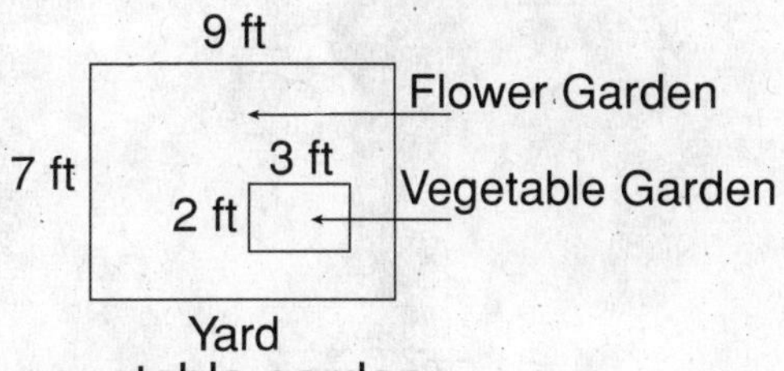

First find the area of the yard and the area of the vegetable garden.

Area of yard: $A = lw = (9 \cdot 7) = 63$
Area of vegetable garden: $A = lw = (3 \cdot 2) = 6$

Then subtract the area of the vegetable garden from the area of the yard to find the area of the flower garden.

$63 - 6 = 57$
So the area of the flower garden is 57 square feet.

Solve.

7. A rectangular yard is made up of a rectangular grass lawn surrounding a rectangular patio. The yard is 12 ft by 10 ft. The patio is 5 ft by 4 ft. What is the area of the grass lawn?

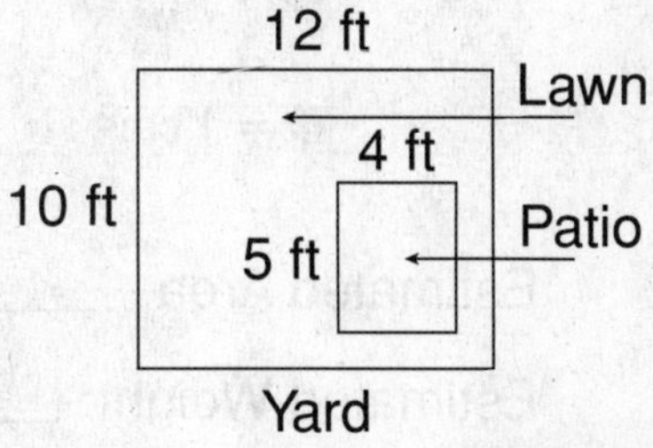

____

Name ______________________ Date ____________ Class ____________

LESSON 10-1

Challenge

Chewing Gum Archaeology

Archaeologists found the oldest piece of chewing gum in Sweden—it was 9,000 years old! The scientists studied the gum to learn about the people who may have made and chewed it. You can be a chewing gum archaeologist, too. How? When a standard-sized wad of chewed gum is stepped on, you can study the area of the flattened gum to find the weight of the person who squished it!

Estimate the area covered by each flattened piece of gum below. Then use the table to find the gum squisher's weight.

Bubble gum was invented in Pennsylvania in 1928.

Estimated Area of Squished Gum	Estimated Weight of Gum Squisher
about 25 cm^2	about 50 lb
about 50 cm^2	about 100 lb
about 75 cm^2	about 150 lb
about 100 cm^2	about 200 lb

1.

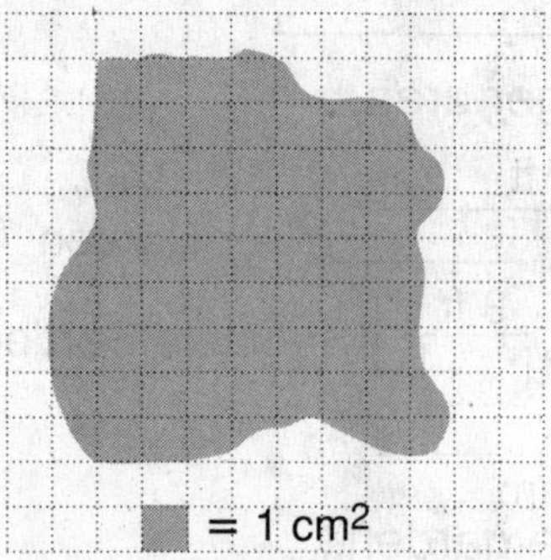

Estimated Area: ____________

Estimated Weight: ____________

2.

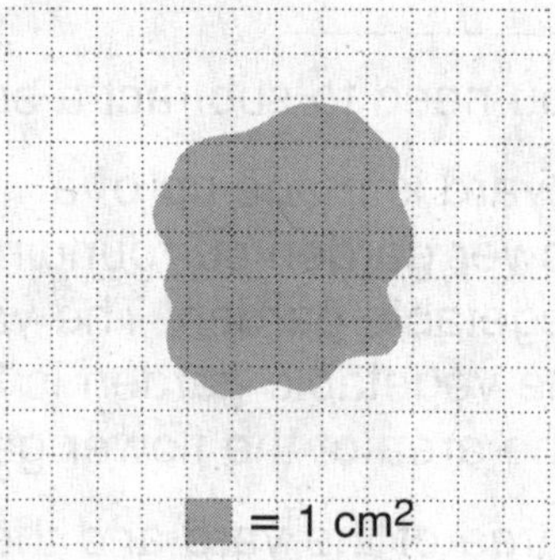

Estimated Area: ____________

Estimated Weight: ____________

3.

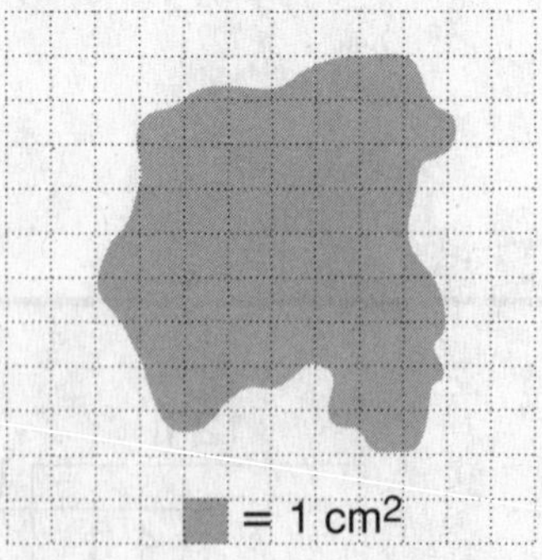

Estimated Area: ____________

Estimated Weight: ____________

4.

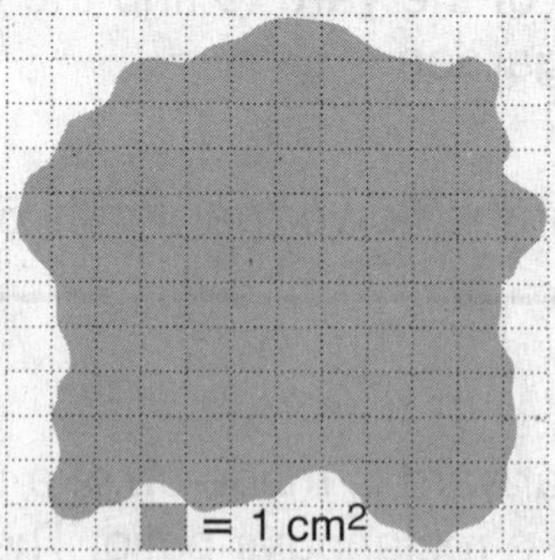

Estimated Area: ____________

Estimated Weight: ____________

Name ______________________ Date __________ Class __________

LESSON 10-1

Problem Solving

Estimating and Finding Area

Use the table to answer each question.

State Information

State	**Approx. Width** (mi)	**Approx. length** (mi)	**Water Area** (mi^2)
Colorado	280	380	376
Kansas	210	400	462
New Mexico	343	370	234
North Dakota	211	340	1,724
Pennsylvania	160	283	1,239

1. New Mexico is the 5th largest state in the United States. What is its approximate total area?

____

2. Kansas is the 15th largest state in the United States. What is its approximate total area?

____

3. What is the difference between North Dakota's land area and water area?

____

4. What is Pennsylvania's approximate land area?

____

Circle the letter of the correct answer.

5. What is the difference between Colorado's land area and Pennsylvania's land area?

A 106,400 mi^2

B 61,120 mi^2

C 60,120 mi^2

D 45,280 mi^2

6. About what percent of the total area of Pennsylvania is covered by land?

F about 3%

G about 30%

H about 67%

J about 97%

7. Rhode Island is the smallest state. Its total land area is approximately 1,200 mi^2. Rhode Island is approximately 40 miles long. About how wide is Rhode Island?

A about 20 mi

B about 40 mi

C about 50 mi

D about 30 mi

8. The entire United States covers 3,794,085 square miles of North America. About how much of that area is not made up of the 5 states in the chart?

F 2,537,470 mi^2

G 3,359,755 mi^2

H 3,686,525 mi^2

J 3,1310,818 mi^2

Name ______________________ Date ____________ Class ____________

LESSON 10-1

Reading Strategies

Use a Flowchart

Area is the measure of the amount of surface a figure covers.

Area is measured in square units.

This pond has a grid placed over it so that the number of square units the pond covers can be estimated.

Follow this flowchart to help you estimate the area of the pond.

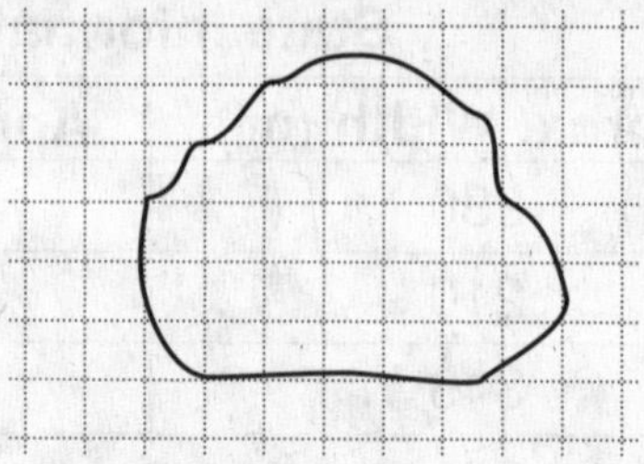

Step 1: Count full squares and almost full squares.

↓

Step 2: Count squares that are about half full. Add two half squares together to make a full square.

↓

Step 3: Do not count squares that are almost empty.

↓

Step 4: Add to estimate the total.

Answer the following questions.

1. What do you call the amount of surface a figure covers?

____

2. What is the first step in estimating the area of the pond above?

____

3. What is your estimate for the number of full squares?

____

4. What is the second step in estimating the area of the pond?

____

5. About how many half squares did you count?

____

6. What is your estimate for the area of the pond?

____

Name ____________________ Date __________ Class __________

LESSON **10-1**

Puzzles, Twisters & Teasers

What's Furry?

Find the area of each figure below. Then find the letter matching the answer in the Answer Box. To solve the riddle, match the letter of the correct answer to the number of the problem.

1.

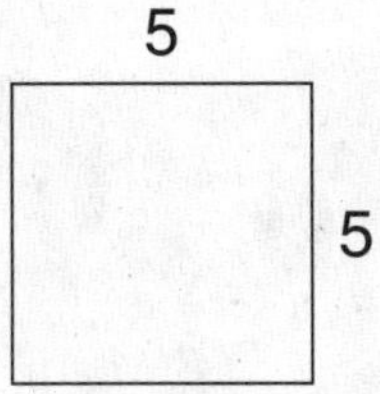

Area = ________ units^2

2.

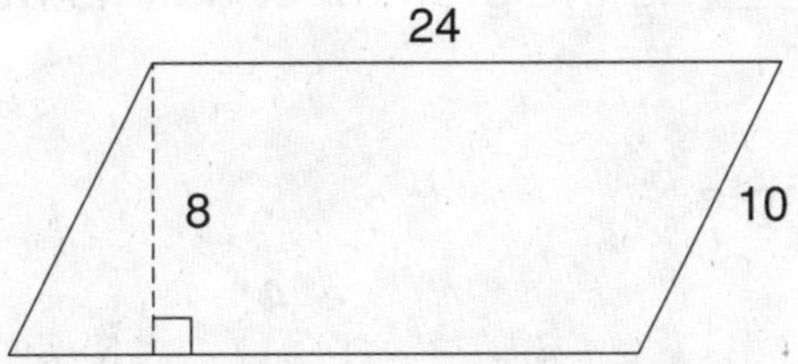

Area = ________ units^2

3.

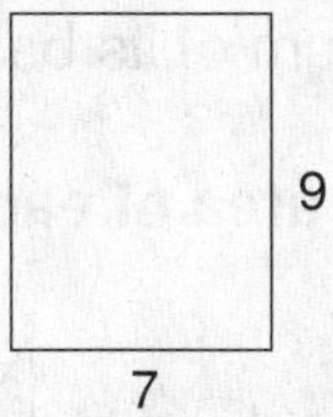

Area = ________ units^2

4.

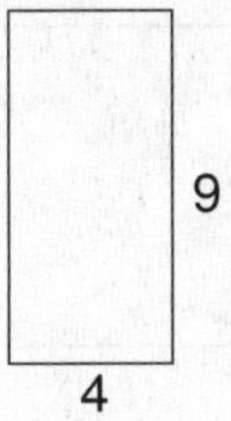

Area = ________ units^2

5.

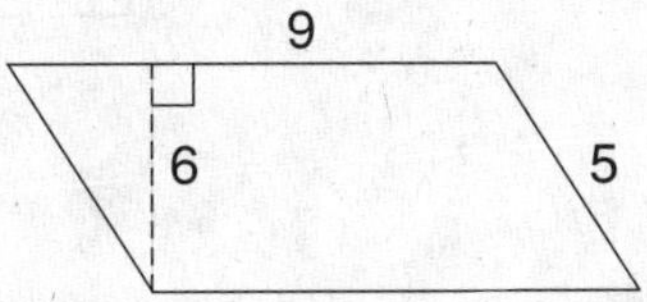

Area = ________ units^2

6.

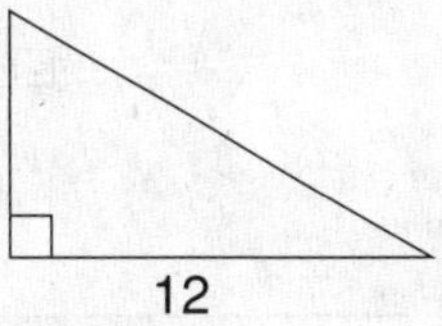

Area = ________ units^2

7.

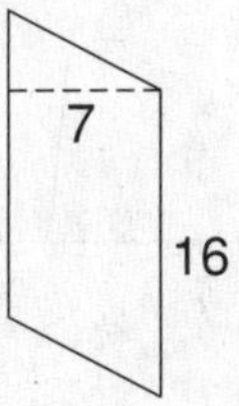

Area = ________ units^2

8.

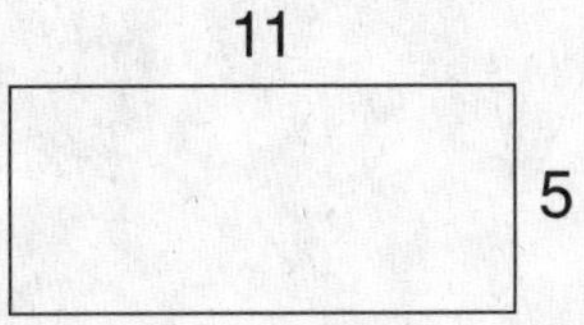

Area = ________ units^2

9.

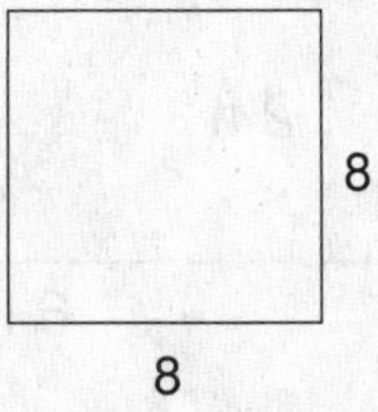

Area = ________ units^2

Answer Box

A = 112	**D** = 26	**H** = 42	**K** = 64	**N** = 55	**R** = 15
B = 28	**E** = 46	**I** = 63	**L** = 110	**O** = 54	**T** = 26
C = 25	**G** = 30	**J** = 240	**M** = 192	**P** = 36	**W** = 72

What do you call a monkey that sells potato chips?

____ ____ ____ ____ ____ ____ ____ ____ ____

7 1 6 3 4 2 5 8 9

Name ______________________ Date ____________ Class ____________

LESSON **10-2**

Practice A

Area of Triangles and Trapezoids

Complete. Write *triangle* or *trapezoid*.

1. The area of a ____________ is half the product of its base and its height.

2. The area of a ____________ is the product of half its height and the sum of its bases.

Find the area of each triangle.

3.

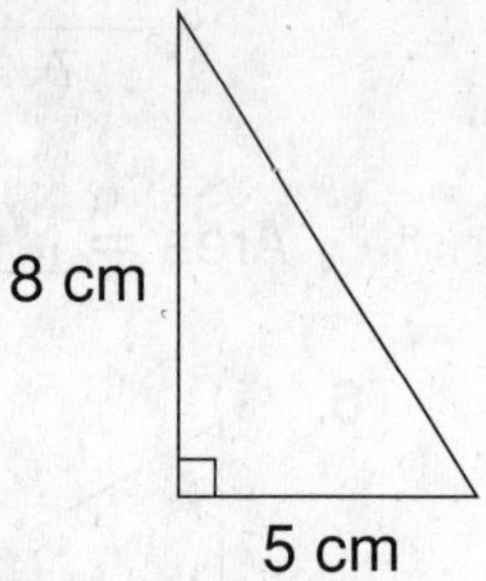

____

4.

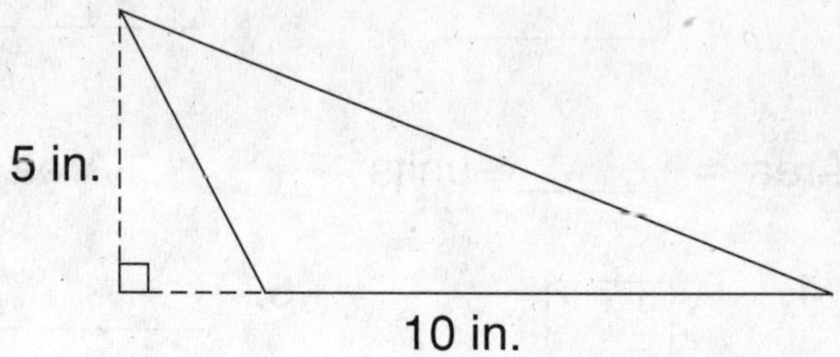

____

Find the area of each trapezoid.

5.

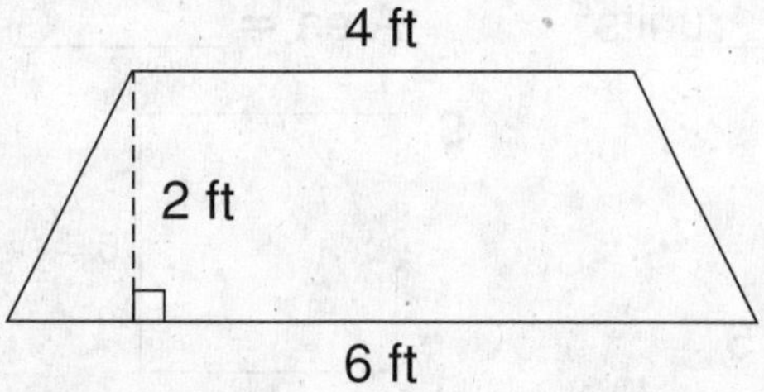

____

6.

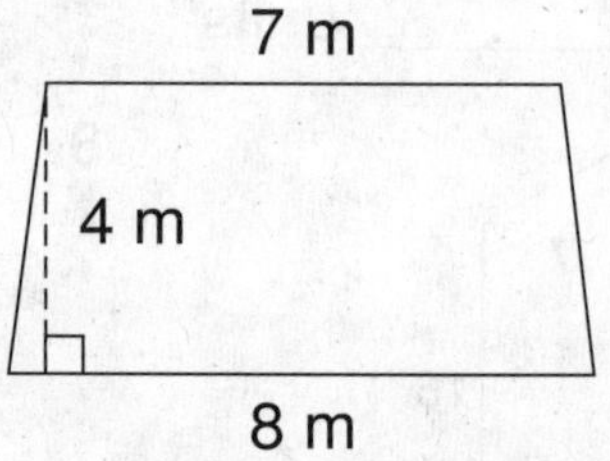

____

7. Andrew is making a triangular jewelry box. What is the area of the top of the box?

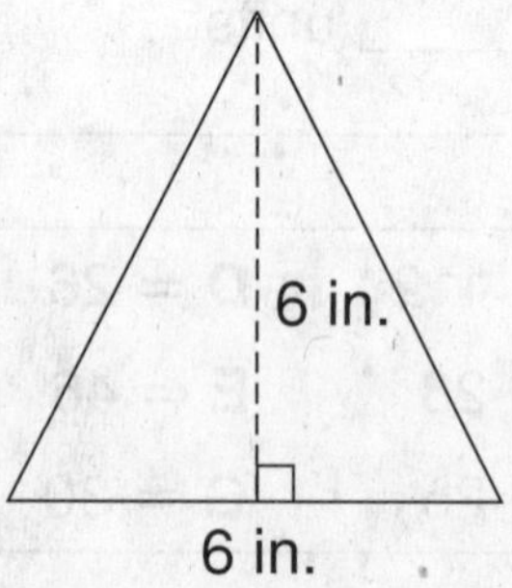

____

Name ______________________ Date __________ Class __________

LESSON **10-2**

Practice B

Area of Triangles and Trapezoids

Find the area of each triangle.

1.

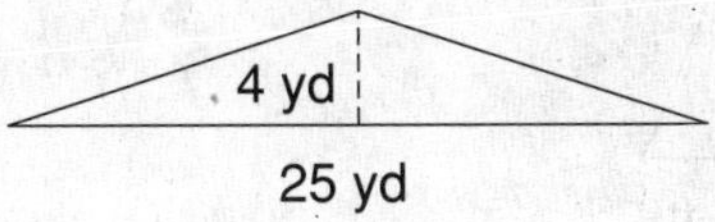

____

2.

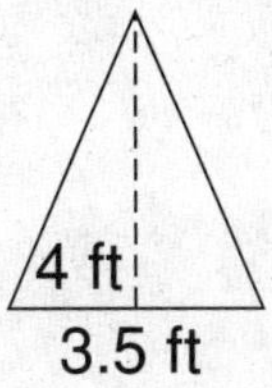

____

3.

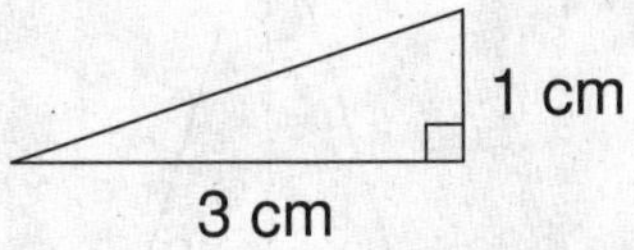

____

4.

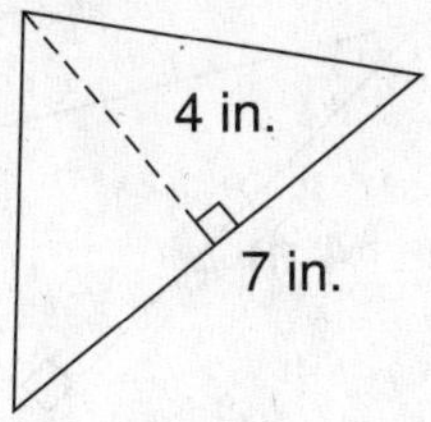

____

Find the area of each trapezoid.

5.

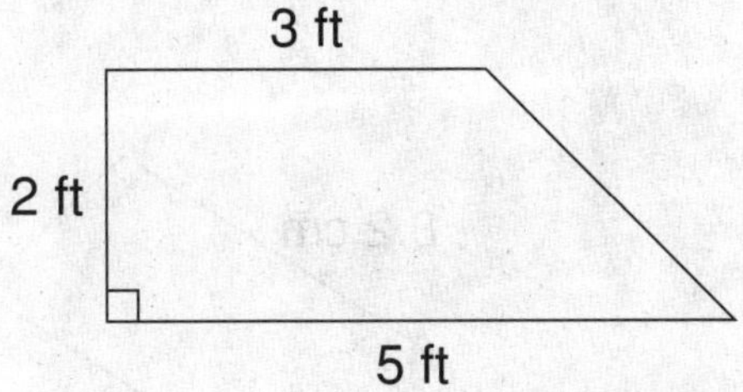

____

6.

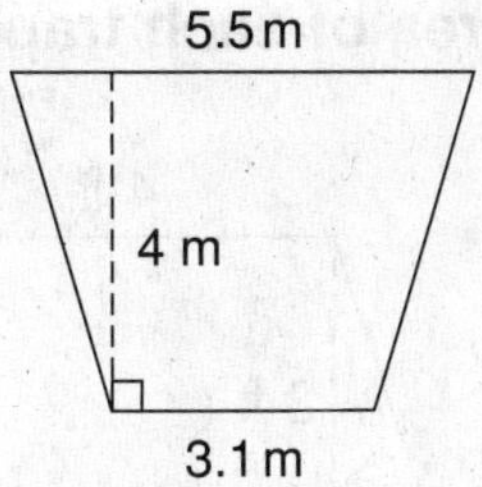

____

7.

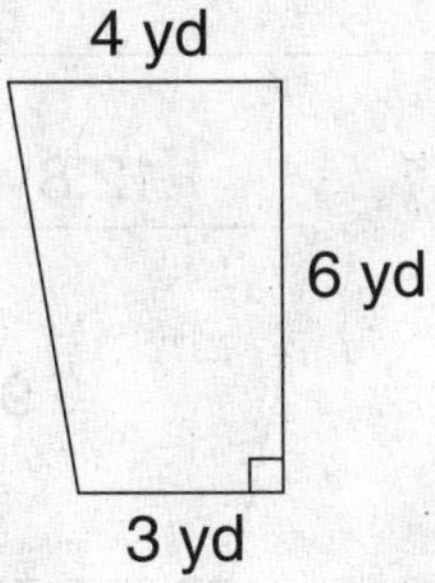

____

8.

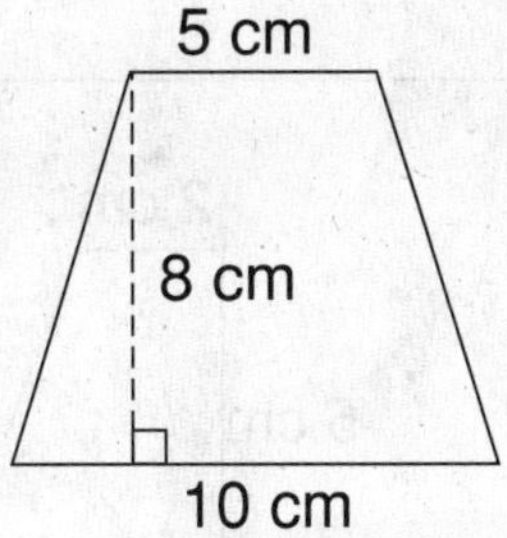

____

9. The front part of a tent is 8 feet long and 5 feet tall. What is the area of the front part of the tent?

____

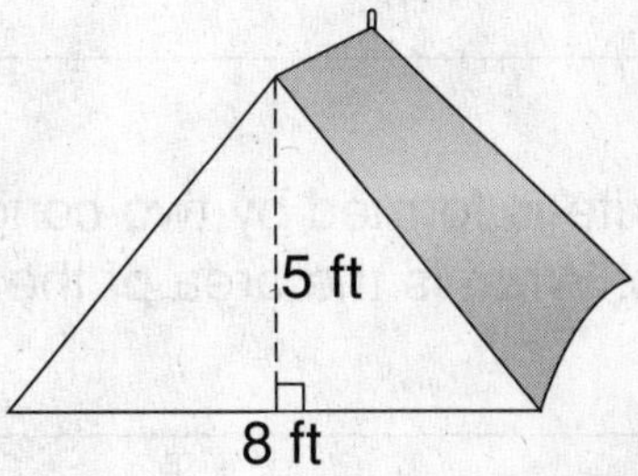

Name ______________________ Date ____________ Class ____________

LESSON **10-2**

Practice C

Area of Triangles and Trapezoids

Find the area of each triangle.

1.

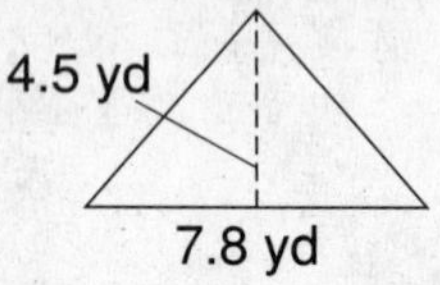

____

2.

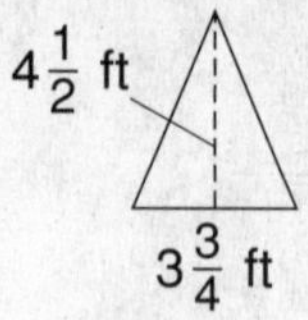

____

3.

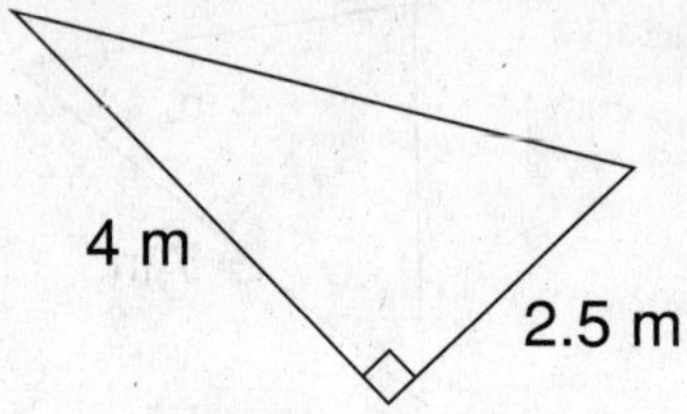

____

4.

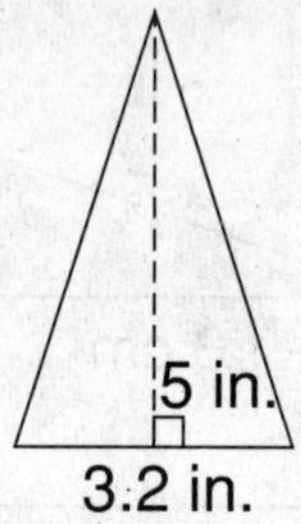

____

Find the area of each trapezoid.

5.

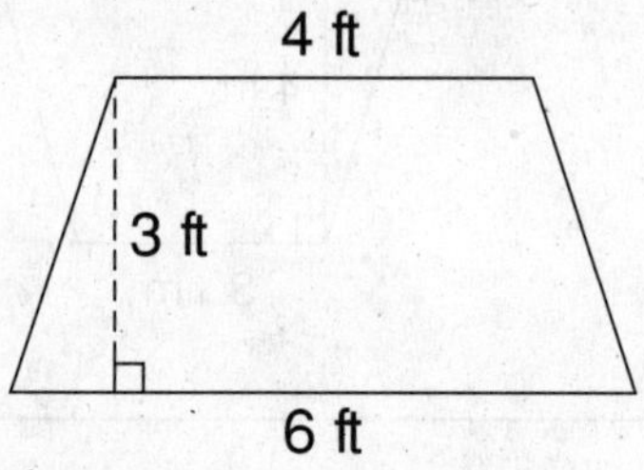

____

6.

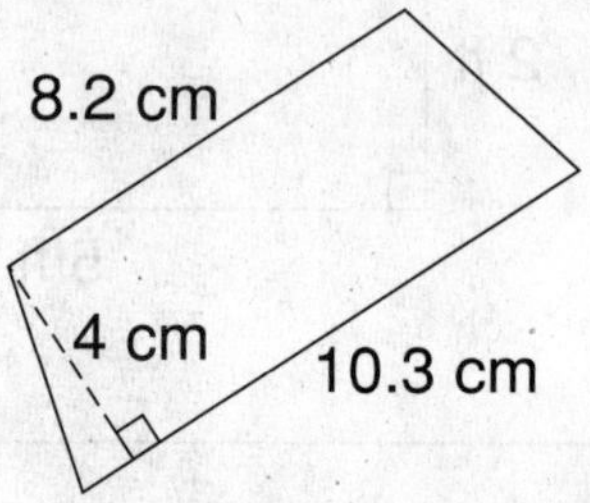

____

7.

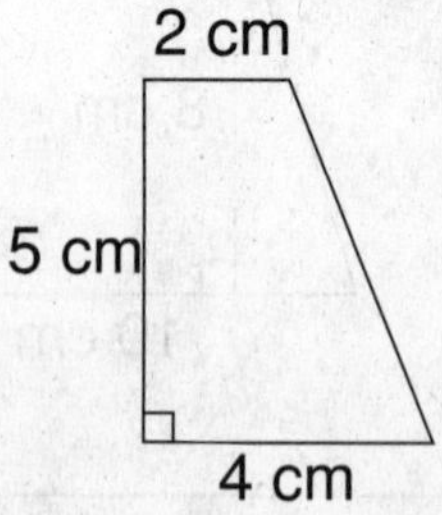

____

8.

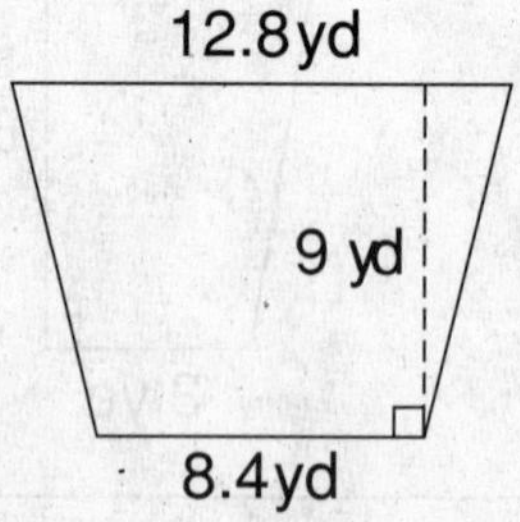

____

9. A kite is formed by two congruent triangles with 6.2-cm bases and heights of 8.5 cm. What is the area of the entire kite?

____

Name ______________________ Date __________ Class __________

LESSON **10-2**
Reteach
Area of Triangles and Trapezoids

To find the area of a triangle, first turn your triangle into a rectangle.

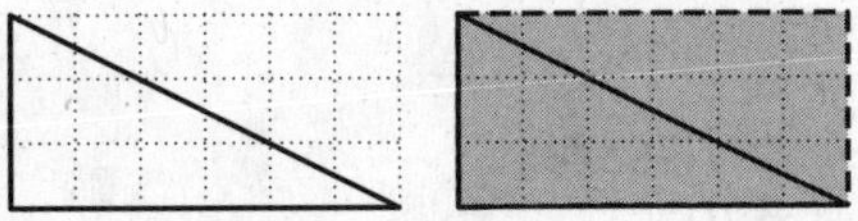

Next, find the area of the rectangle. 6 • 3 = 18

The triangle is half the area of the formed rectangle or $A = \frac{1}{2}bh$, so divide the product by 2.

$18 \div 2 = 9$ So, the area of the triangle is 9 square units.

Find the area of each triangle.

1.

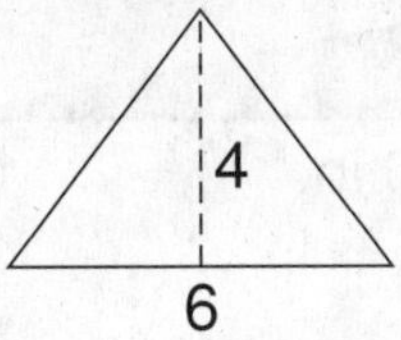

____

2.

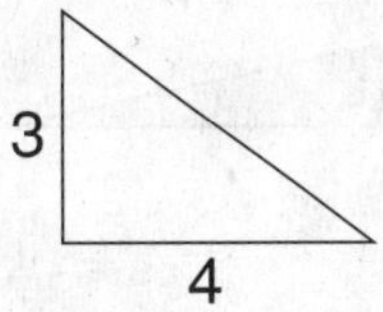

____

To find the area of a trapezoid, first turn the trapezoid into two triangles and a rectangle.

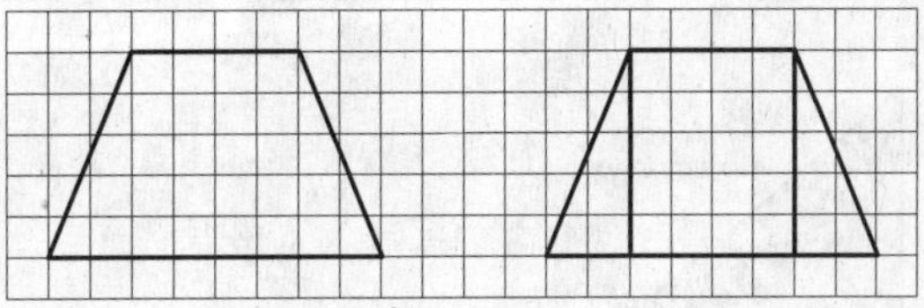

Find the area of each triangle.

$A = \frac{1}{2}bh$

$A = \frac{1}{2}(2 \cdot 5) = \frac{1}{2} \cdot 10 = 5$

Find the area of the rectangle.

$A = lw$

$A = 4 \cdot 5 = 20$

Now find the sum of the areas.

$5 + 5 + 20 = 30$ So, the area of the trapezoid is 30 square units.

Find the area of each trapezoid.

3.

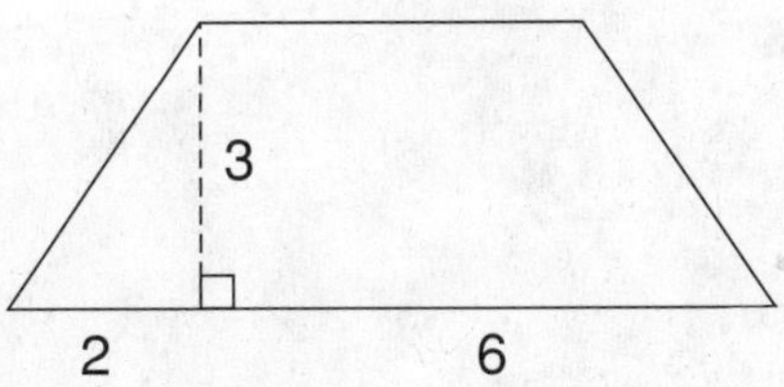

____

4.

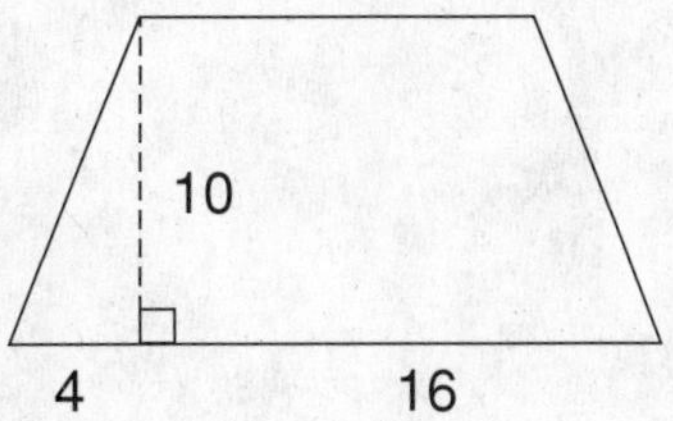

____

Name ______________________ Date ____________ Class ____________

LESSON 10-2

Challenge

Pennant Problems

Find the amount of felt used to make each pennant.

1.

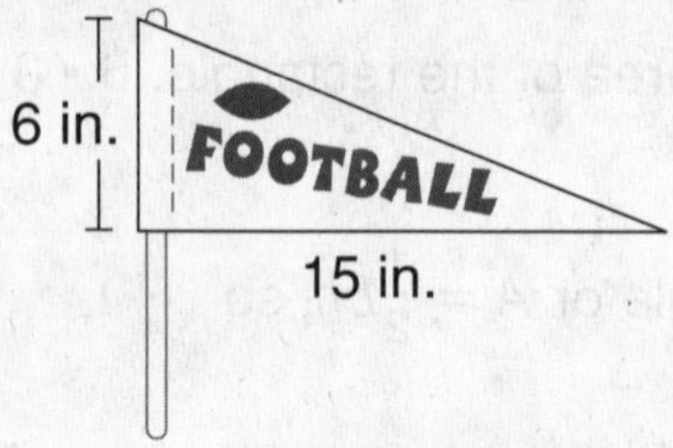

____

2.

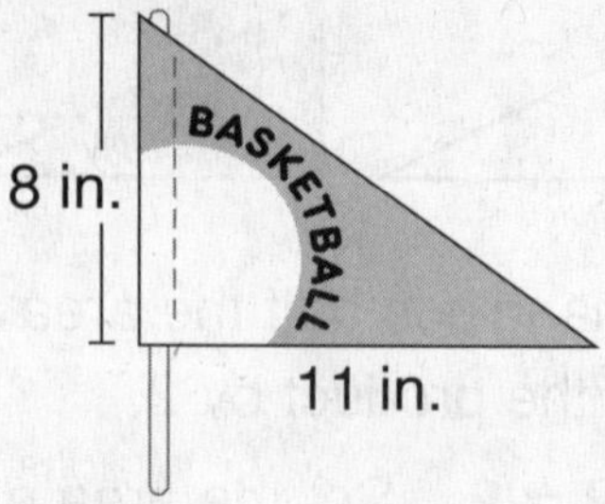

____

3.

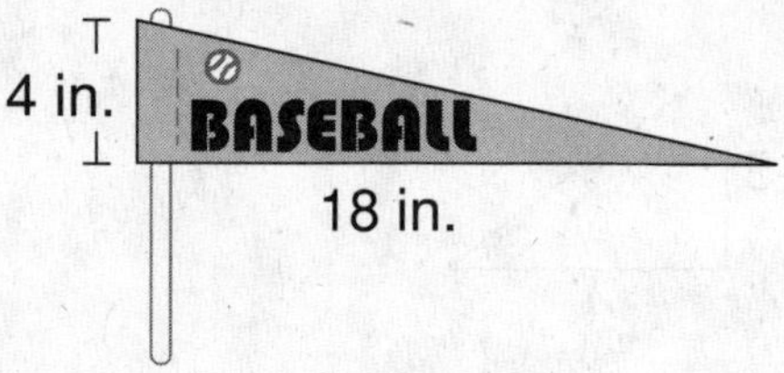

____

4.

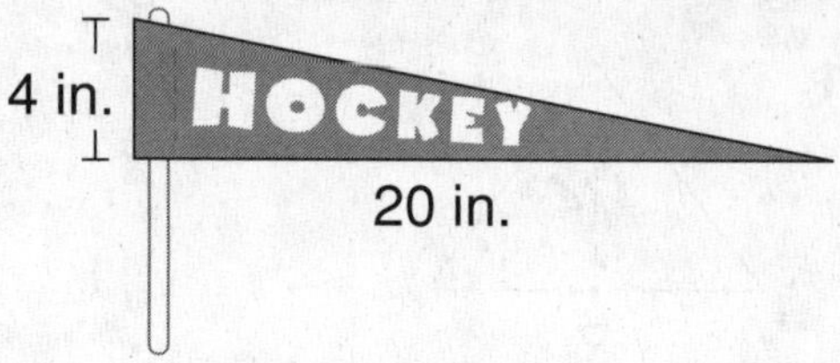

____

5.

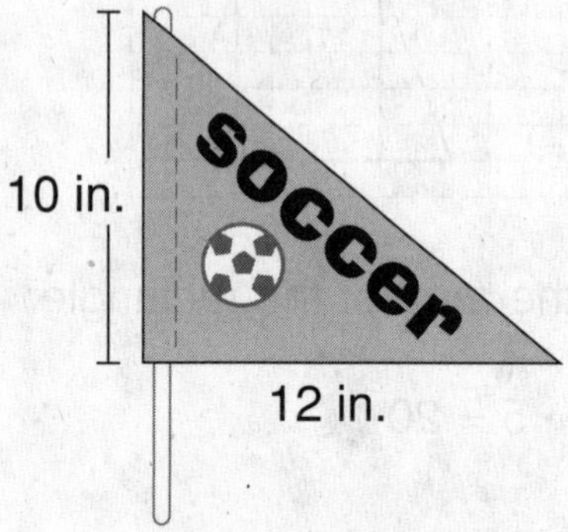

____

6.

____

7. List the pennants in order from least area to greatest area.

____

Name ______________________ Date __________ Class __________

LESSON **10-2**

Problem Solving

Area of Triangles and Trapezoids

Use the quilt design to answer the questions.

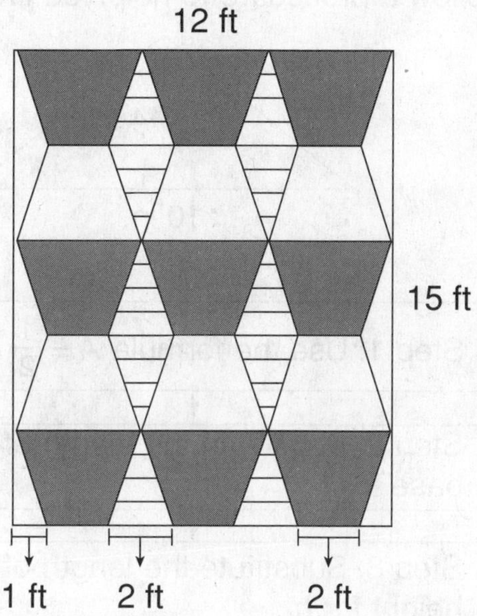

1. What are the lengths of the bases of each trapezoid?

____

2. What is the height of each trapezoid?

____

3. What is the area of each trapezoid?

____

Circle the letter of the correct answer.

4. What is the length of the base of each striped triangle?

A 1 ft

B 2 ft

C 3 ft

D 4 ft

5. What is the height of each striped triangle?

F 1 ft

G 2 ft

H 3 ft

J 5 ft

6. What is the area of each striped triangle?

A 3 ft^2

B 1 ft^2

C $\frac{3}{4}$ ft^2

D $\frac{1}{4}$ ft^2

7. What is the area of the quilt?

F 36 ft^2

G 90 ft^2

H 96 ft^2

J 180 ft^2

Name ______________________ Date ____________ Class ____________

LESSON **10-2**

Reading Strategies

Follow a Procedure

Triangles and trapezoids are two different types of polygons. You can follow a procedure to help you find the area of each type of polygon.

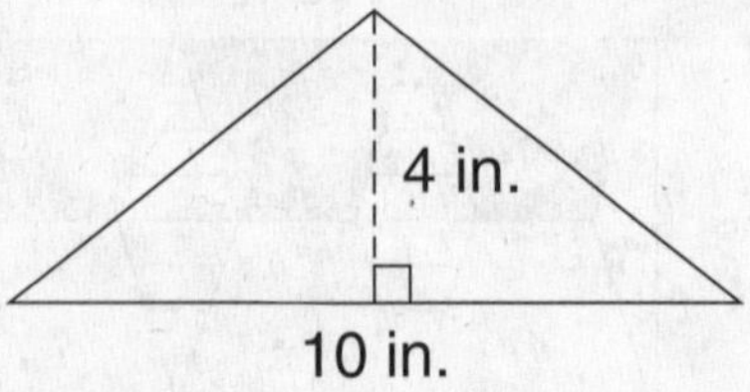

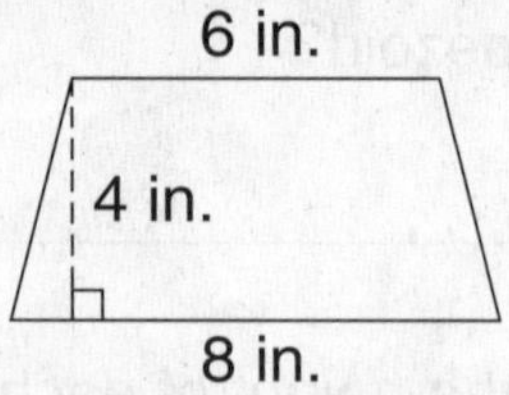

Step 1: Use the formula $A = \frac{1}{2}bh$.

↓

Step 2: Substitute the length of the base for b.

↓

Step 3: Substitute the length of the height for h.

↓

Step 4: Multiply.

Step 1: Use the formula $A = \frac{1}{2}h(b_1 + b_2)$.

↓

Step 2: Substitute the length of the height for h.

↓

Step 3: Substitute the lengths of the bases for b_1 and b_2 and add.

↓

Step 4: Multiply.

Answer each question.

1. What is the first step in finding the area of the triangle?

____

2. What are the second and third steps in finding the area of the triangle?

____

3. What is the area of the triangle?

____

4. What is the first step in finding the area of the trapezoid?

____

5. What are the second and third steps in finding the area of the trapezoid?

____

6. What is the area of the trapezoid?

____

Name ______________________ Date ____________ Class ____________

LESSON **10-2**

Puzzles, Twisters & Teasers

Hear, Hear!

Find the area of each figure below and circle the letter next to the correct answer. Write the letters over the number of the problem in the corresponding space in the riddle.

1.

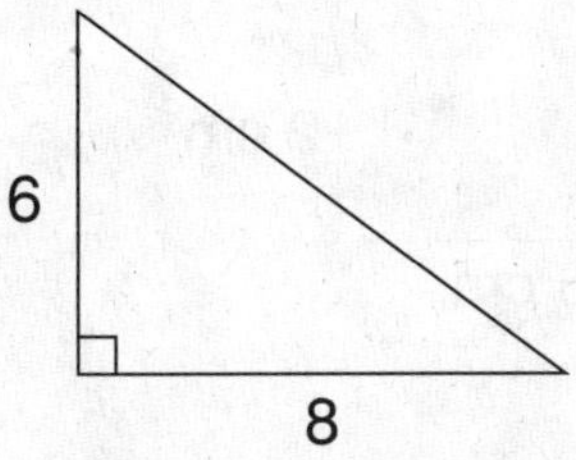

E 24 units² **B** 26 units² **X** 28 units²

2.

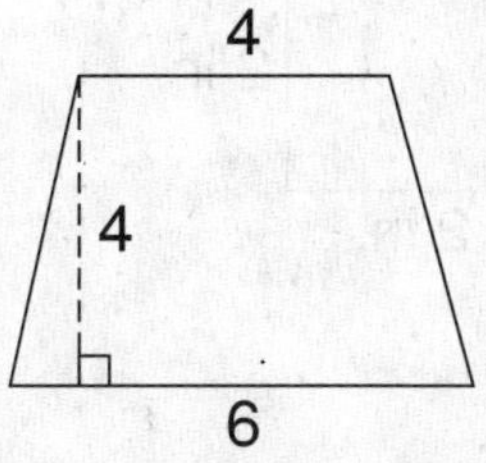

A 20 units² **F** 24 units² **G** 40 units²

3.

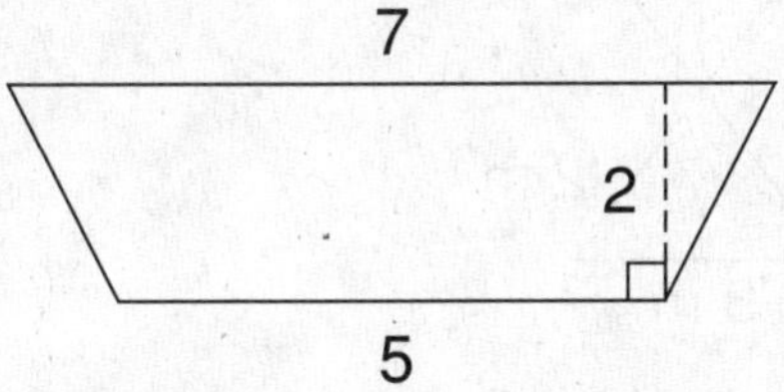

O 10 units² **R** 12 units² **P** 24 units²

4.

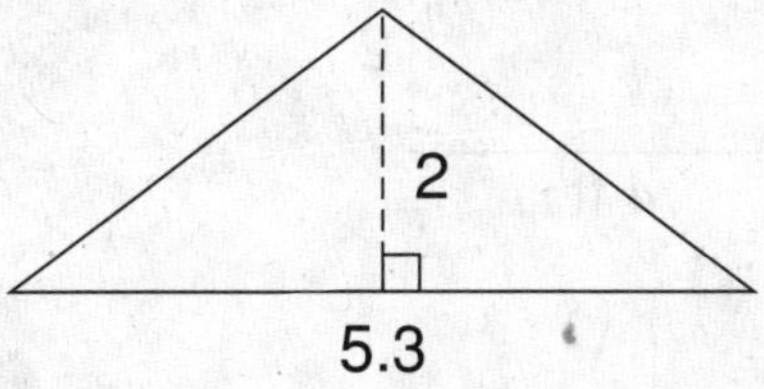

C 5.1 units² **Q** 5.2 units² **D** 5.3 units²

5.

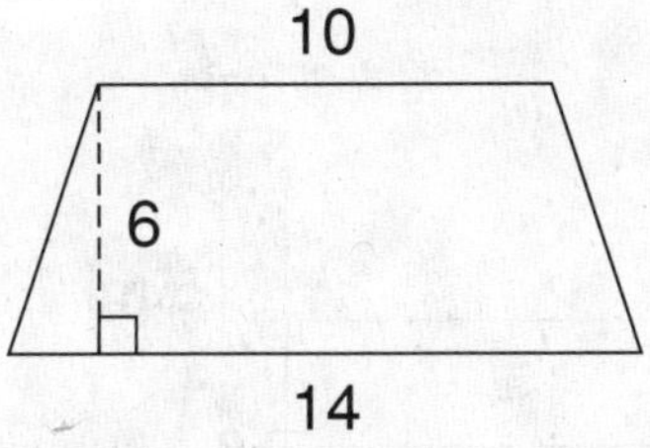

L 24 units² **E** 48 units² **R** 72 units²

6.

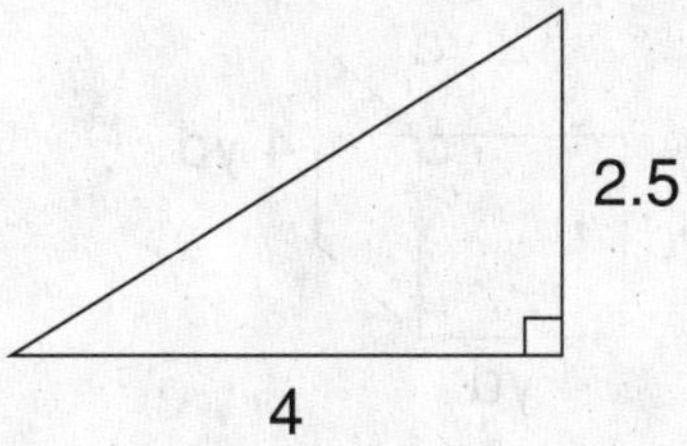

V 4 units² **U** 5 units² **I** 6 units²

7.

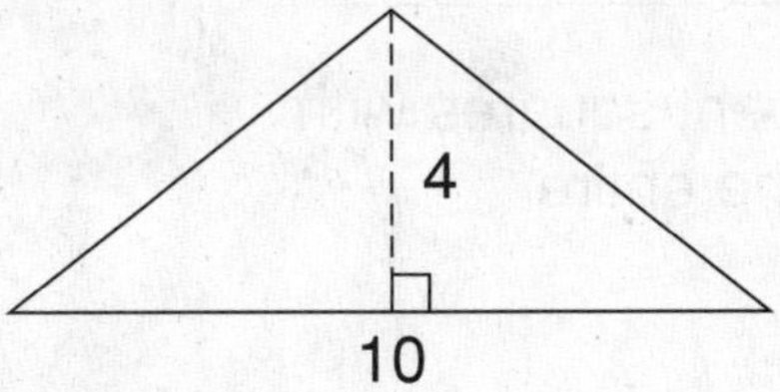

M 20 units² **H** 21 units² **W** 40 units²

8.

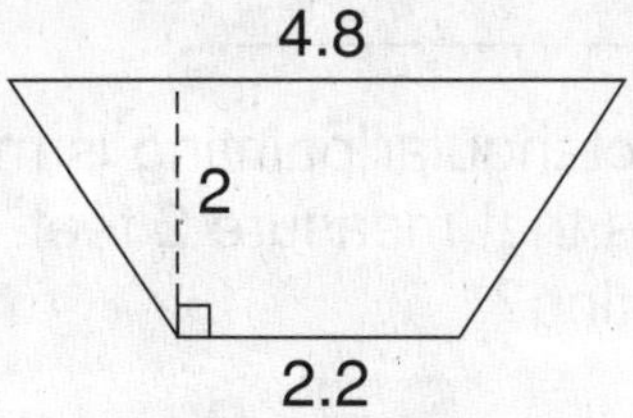

J 5 units² **N** 6 units² **S** 7 units²

What part of your body has the most rhythm?

____ ____ ____ ____ ____ ____ ____ ____

1. 2. 3. 4. 5. 6. 7. 8.

Name ______________________ Date ____________ Class ____________

LESSON **10-3**

Practice A
Area of Composite Figures

Find the area of each polygon.

1.

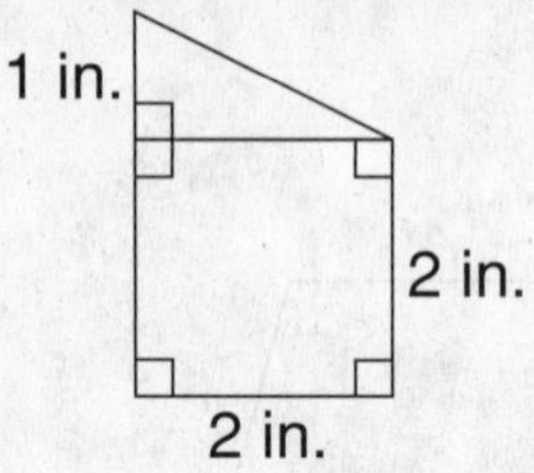

____

2.

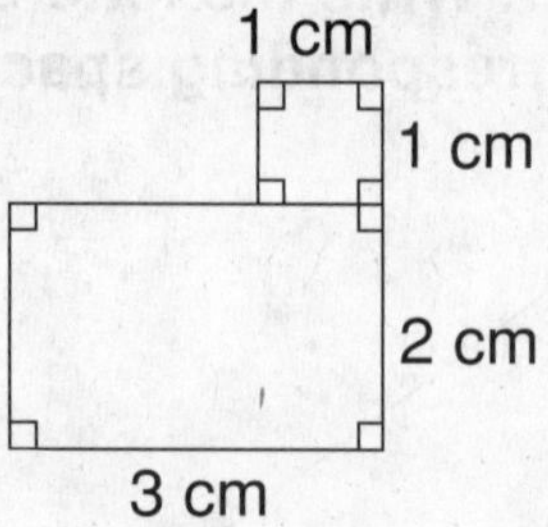

____

3.

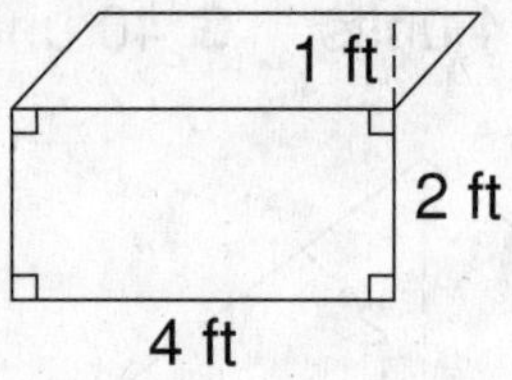

____

4.

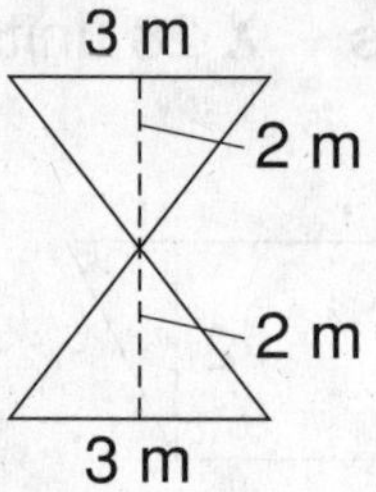

____

5.

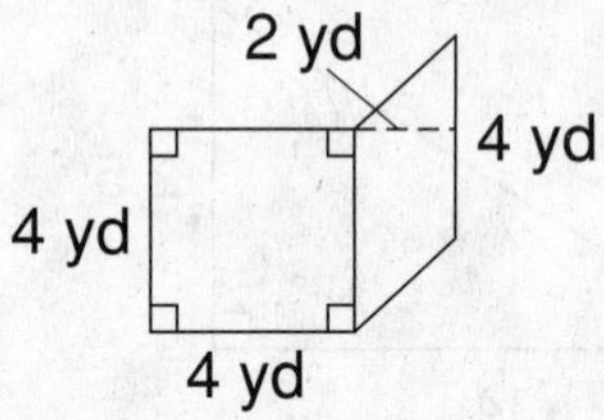

____

6.

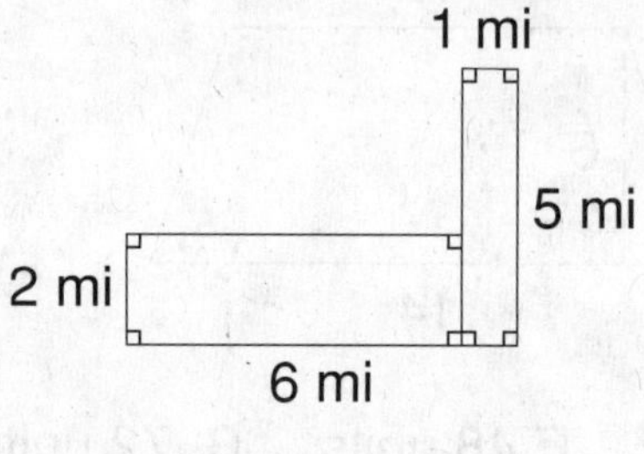

____

7. A rectangular painting is made up of two congruent squares with sides that measure 2 feet. What is the area of the entire painting?

____

8. A carpet is made up of two congruent triangles. The base of each triangle is 3 feet and the height is 6 feet. What is the area of the entire carpet?

____

Name ______________________________ Date __________ Class __________

LESSON 10-3

Practice B
Area of Composite Figures

Find the area of each polygon.

1.

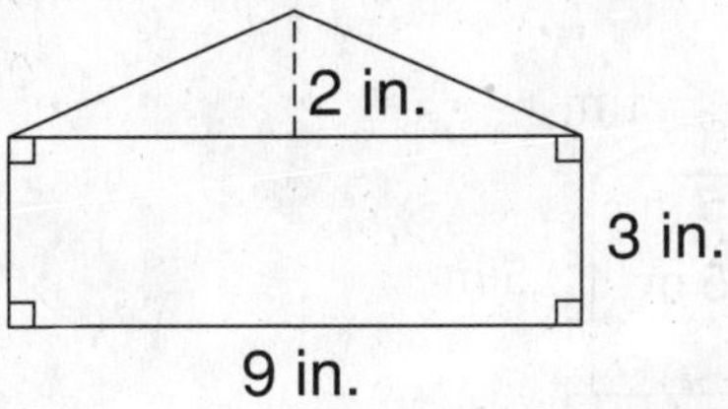

____

2.

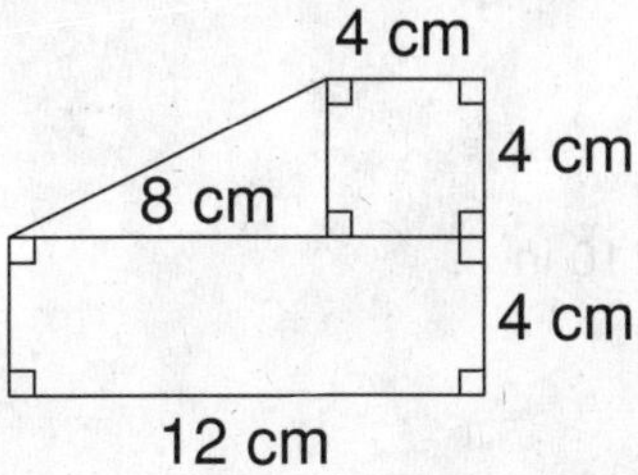

____

3.

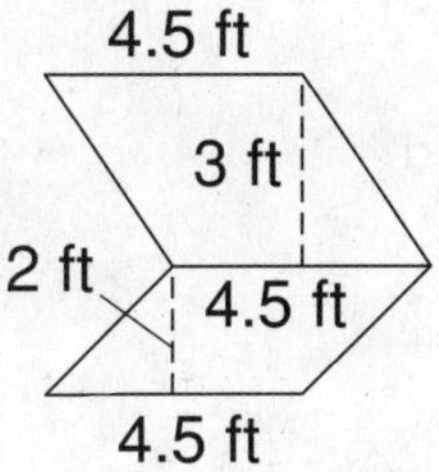

____

4.

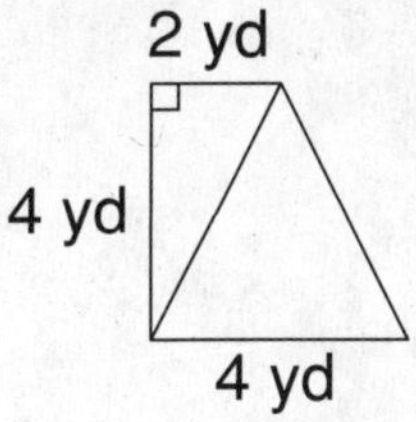

____

5.

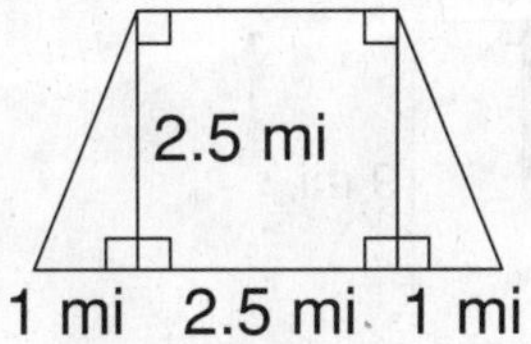

____

6.

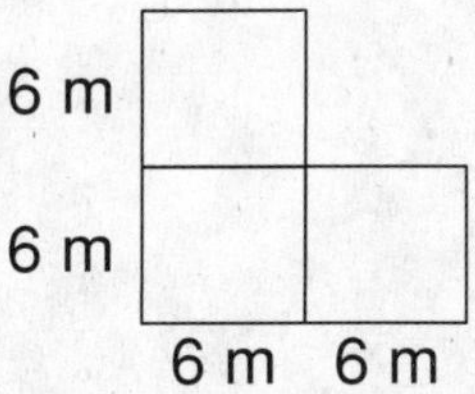

____

7. Three paintings are shaped like an 8-foot square, a 7-foot by 4-foot rectangle, and a triangle with a 6-foot base and a height of 7 feet. If those paintings are hung together on the outside of a building, how much of the building's wall will they cover altogether?

____

8. Two diagonals divide a square carpet into 4 congruent triangles. The base of each triangle is 5 feet and the height is 2.5 feet. What is the area of the entire carpet?

____

Name ______________________ Date __________ Class __________

LESSON 10-3

Practice C
Area of Composite Figures

Find the area of each figure. The shaded parts are cut out of the figures.

1.

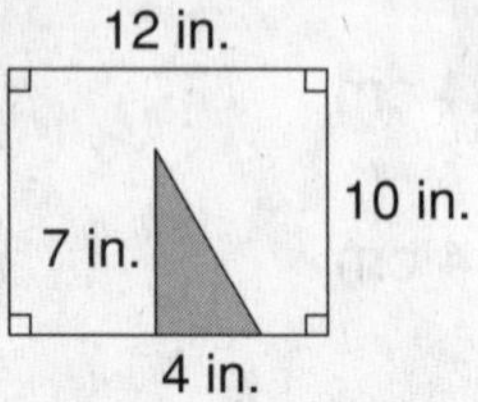

____

2.

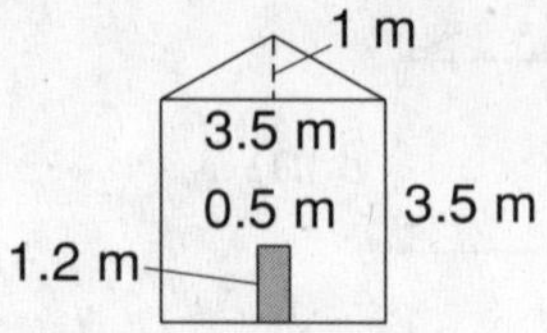

____

3.

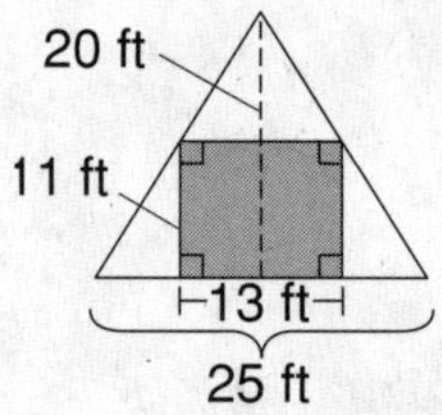

____

4.

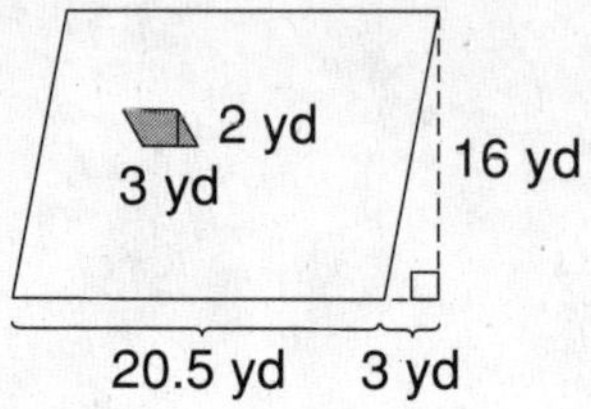

____

5.

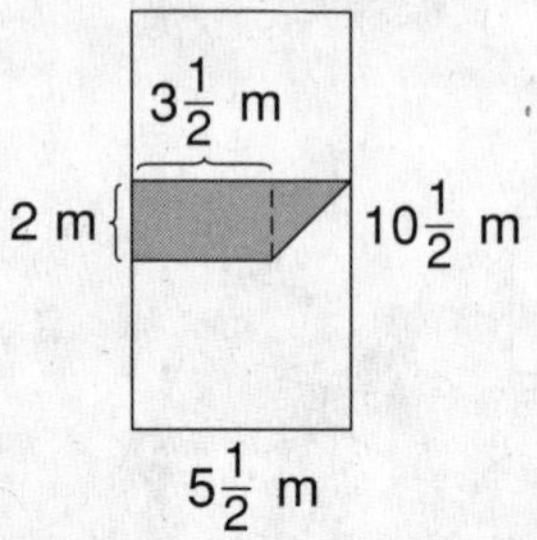

____

6.

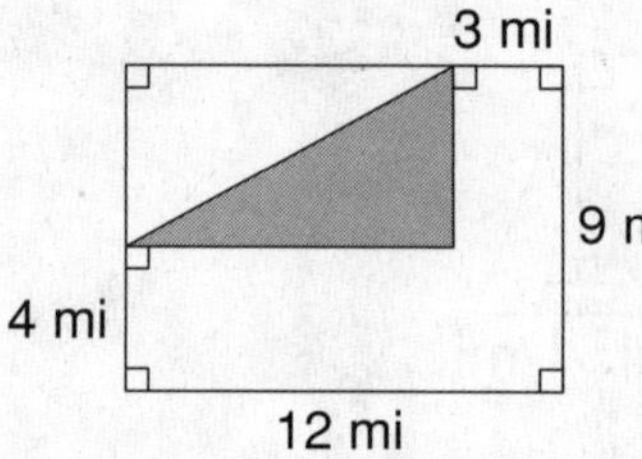

____

7. The front wall of the shed that Rita built is 9 feet tall and $5\frac{1}{4}$ feet wide. After she cut a section out for a door, the area of that wall was $26\frac{1}{4}$ square feet. If the measurements of the doorway were whole numbers of feet, what size door did Rita install?

____

8. The perimeter of a rectangular playground is 40 yards. It is $1\frac{1}{2}$ yards longer than it is wide. What is the area of that playground?

____

Name ______________________________ Date __________ Class __________

LESSON **10-3**

Reteach

Area of Composite Figures

Sometimes you can use area formulas you know to help you find the area of other figures.

To find the area of the figure below, first divide the figure into figures you know.

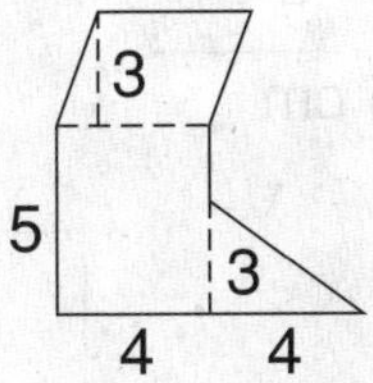

The figure is made up of a triangle, a parallelogram, and a rectangle.

Next, find the area of each figure.

Triangle	Parallelogram	Rectangle
$A = \frac{1}{2}bh$	$A = bh$	$A = \ell w$
$= \frac{1}{2}(3 \cdot 4)$	$= 3 \cdot 4$	$= 4 \cdot 5$
$= 6$	$= 12$	$= 20$

Then, find the sum of all of the areas.

$6 + 12 + 20 = 38$ The area of the figure is 38 square units.

Find the area of each figure.

1.

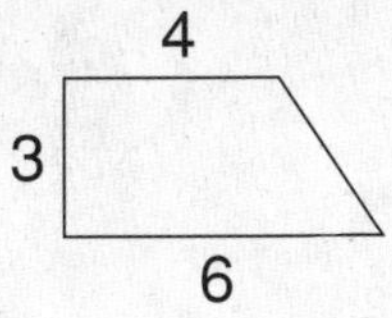

____

2.

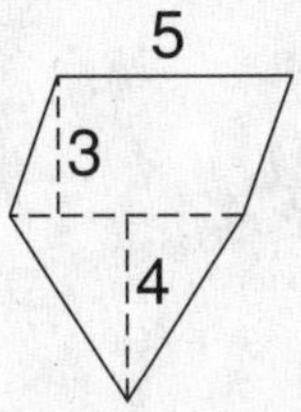

____

3.

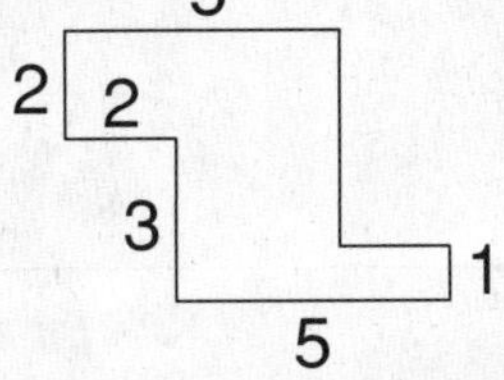

____

4.

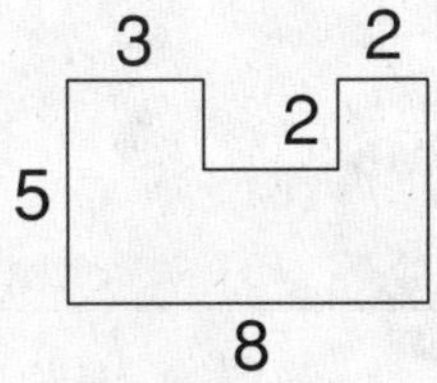

____

Name ______________________ Date ____________ Class ____________

LESSON 10-3

Challenge

Area Combinations

Combine these polygons to draw figures with the given areas. You must use all the polygons, and each can be used only once.

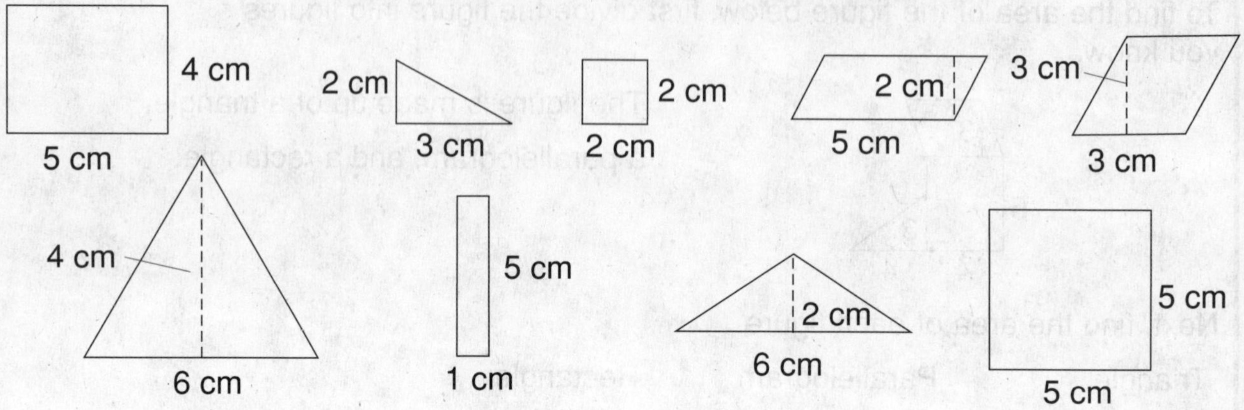

1. Total Area = 26 cm^2

____

2. Total Area = 30 cm^2

____

3. Total Area = 22 cm^2

____

4. Total Area = 16 cm^2

____

Name ______________________________ Date ____________ Class ____________

LESSON 10-3

Problem Solving
Area of Composite Figures

Write the correct answer.

1. The shape of Nevada can almost be divided into a perfect rectangle and a perfect triangle. About how many square miles does Nevada cover?

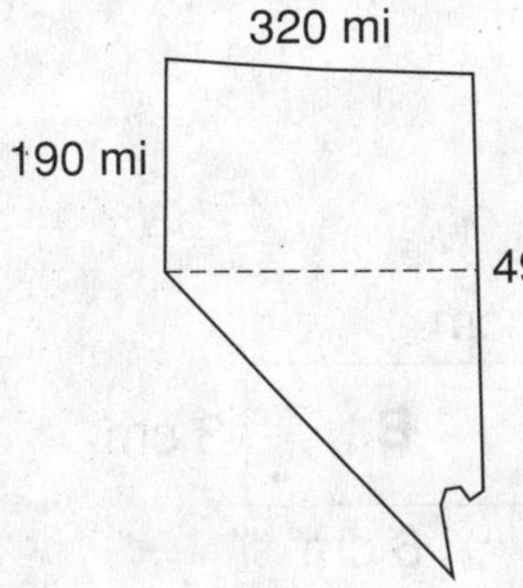

____

2. The shape of Oklahoma can almost be divided into 2 perfect rectangles and 1 triangle. About how many square miles does Oklahoma cover?

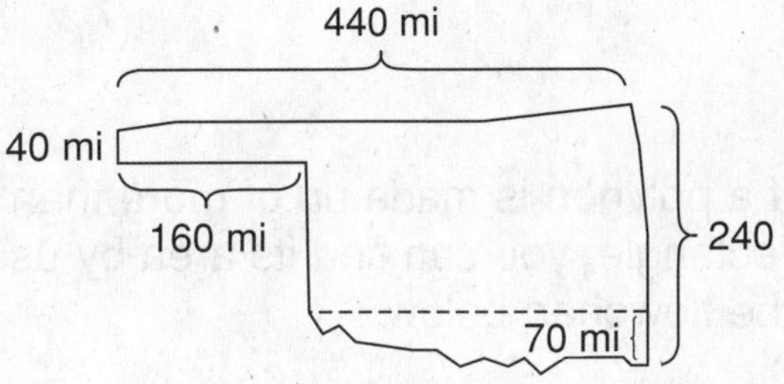

____

3. The front side of an apartment building is a rectangle 60 feet tall and 25 feet wide. Bricks cover its surface, except for a door and 10 windows. The door is 7 feet tall and 3 feet wide. Each window is 4 feet tall and 2 feet wide. How many square feet of bricks cover the front side of the building?

____

____

4. Each side of a square garden is 12 meters long. A hedge wall 1 meter wide surrounds the garden. What is the area of the entire garden including the hedge wall? How many square meters of land does the hedge wall cover alone?

____

____

Circle the letter of the correct answer.

5. A figure is formed by a square and a triangle. Its total area is 32.5 m^2. The area of the triangle is 7.5 m^2. What is the length of each side of the square?

A 5 meters
B 25 meters
C 15 meters
D 16.25 meters

6. A rectangle is formed by two congruent right triangles. The area of each triangle is 6 in^2. If each side of the rectangle is a whole number of inches, which of these could not be its perimeter?

F 26 inches
G 16 inches
H 24 inches
J 14 inches

Name ______________________ Date ____________ Class ____________

LESSON **10-3**

Reading Strategies
Use a Flowchart

To find the area of a rectangle, multiply the length by the width.

$A = \ell \cdot w$

$A = 4$ units $\cdot$ 5 units

$A = 20$ square units

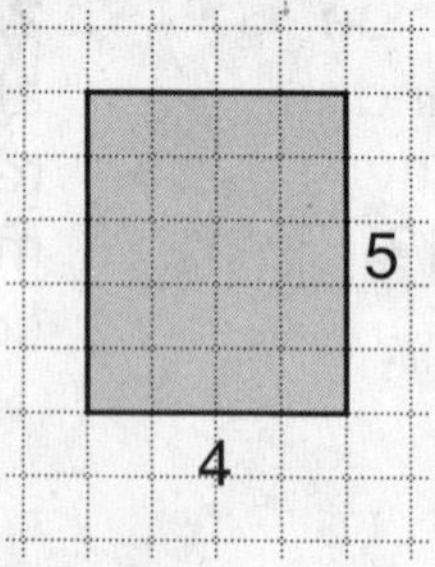

If a polygon is made up of more than one rectangle, you can find its area by using the flowchart below.

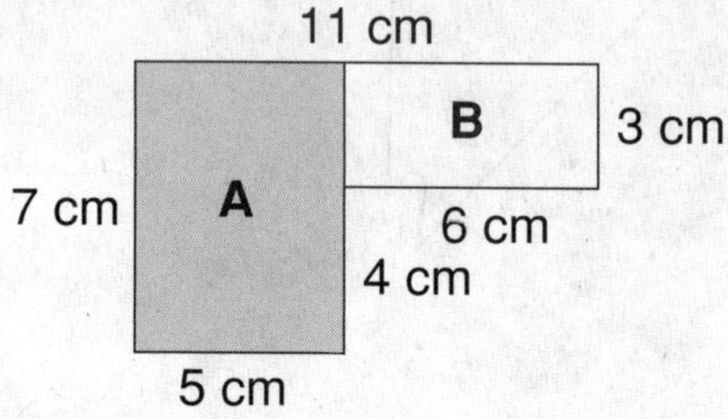

Step 1: Separate the polygon into rectangles.

↓

Step 2: Mark the length and width of each rectangle.

↓

Step 3: Find the area of each rectangle.

↓

Step 4: Add the areas of all of the rectangles together.

Answer the following questions.

1. What is the formula for the area of a rectangle? ____________
2. What are the length and width of rectangle A? ____________
3. What is the area of rectangle A? ____________
4. What is the area of rectangle B? ____________
5. What do you do with the areas of rectangles A and B to find the area of the polygon? ____________
6. What is the area of the polygon? ____________

Name ______________________ Date ____________ Class ____________

LESSON **10-3**

Puzzles, Twisters & Teasers

Sky High

Find the area of each figure. Circle the answers at the bottom of the page. Rearrange the circled letters to solve the riddle.

1.

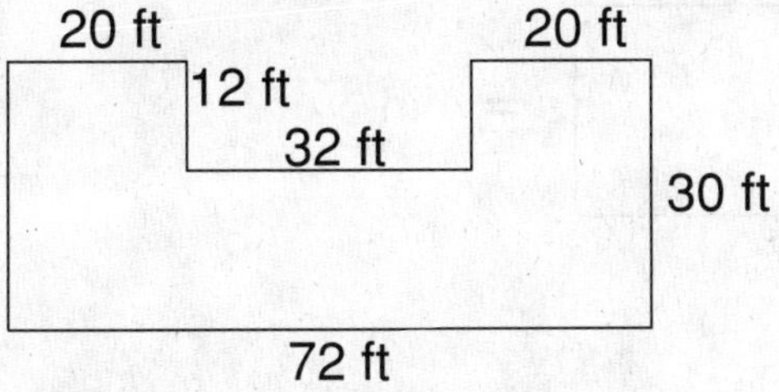

2.

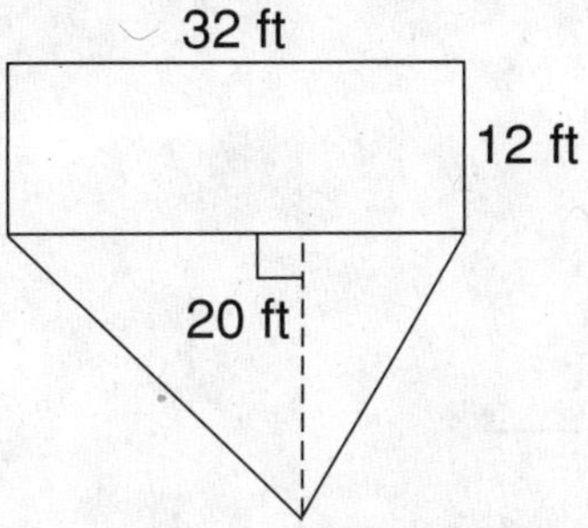

3.

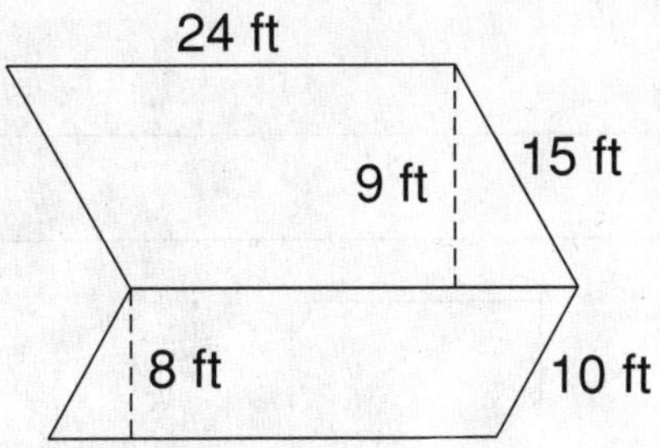

4.

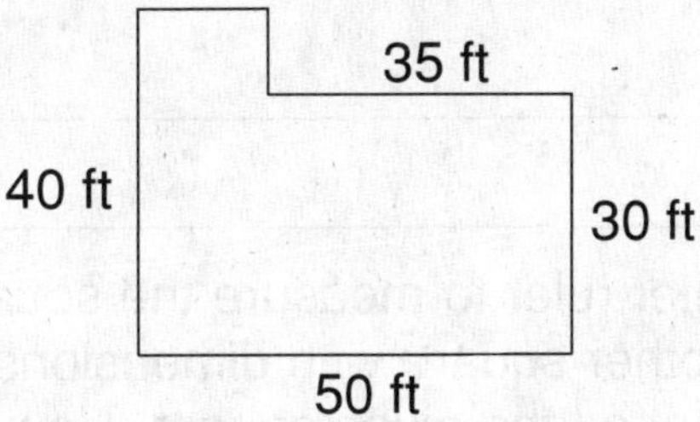

5.

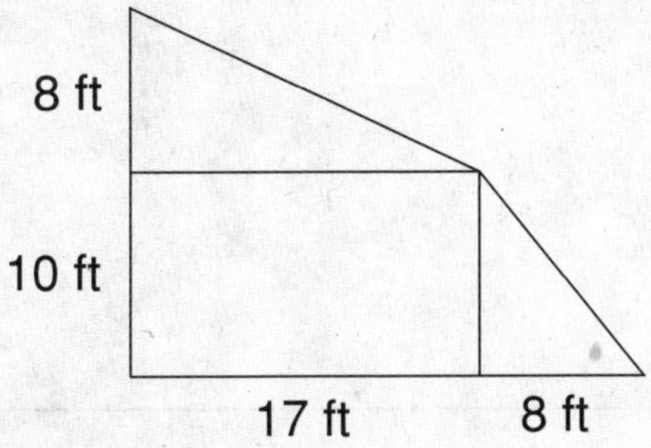

6.

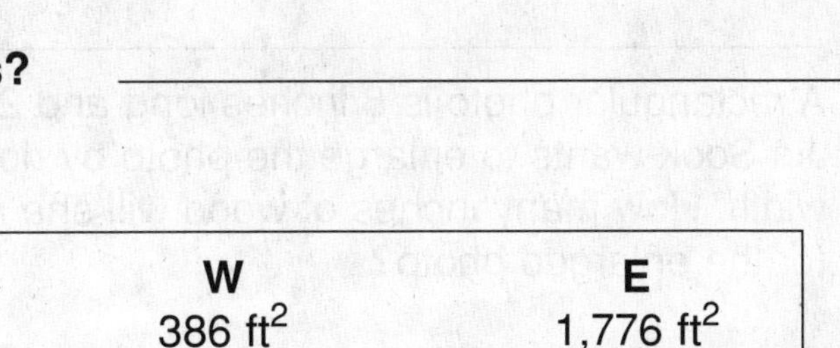

What bird can lift the heaviest weights? ______________________

S 724 ft^2	**A** 408 ft^2	**W** 386 ft^2	**E** 1,776 ft^2
A 704 ft^2	**L** 880 ft^2	**N** 976 ft^2	**I** 1,232 ft^2
M 440 ft^2	**C** 278 ft^2	**T** 1,024 ft^2	**R** 1,650 ft^2

Name ______________________ Date ____________ Class ____________

LESSON **10-4**

Practice A

Comparing Perimeter and Area

Write how the perimeter and the area of the figure change when its dimensions change.

1.  $s = 2$ in.

 $s = 4$ in.

$P =$ ____________ $P =$ ____________

$A =$ ____________ $A =$ ____________

____

____

2. Use a centimeter ruler to measure the square. Then draw another square with dimensions that are half as great as the given square. How do the perimeter and the area change when the dimensions change?

____

____

3. A rectangular photo is 5 inches long and 2 inches wide. Jin-Sook wants to enlarge the photo by doubling its length and width. How many inches of wood will she need to make a frame for the enlarged photo?

____

4. How much glass will Jin-Sook need to cover the enlarged photo from Exercise 3?

____

Name ______________________ Date ____________ Class ____________

LESSON **10-4**

Practice B

Comparing Perimeter and Area

Write how the perimeter and the area of the figure change when its dimensions change.

1. 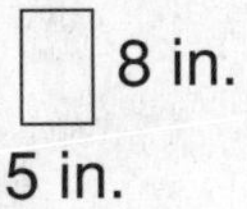

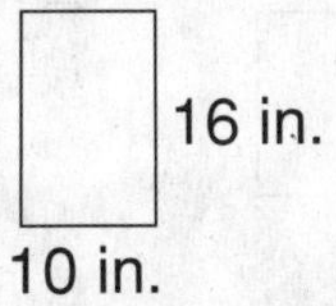

$P =$ ____________ $P =$ ____________

$A =$ ____________ $A =$ ____________

____

____

2. Use a centimeter ruler to measure the triangle. Then draw another triangle with dimensions that are half as great as the given triangle. How do the perimeter and the area change when the dimensions change?

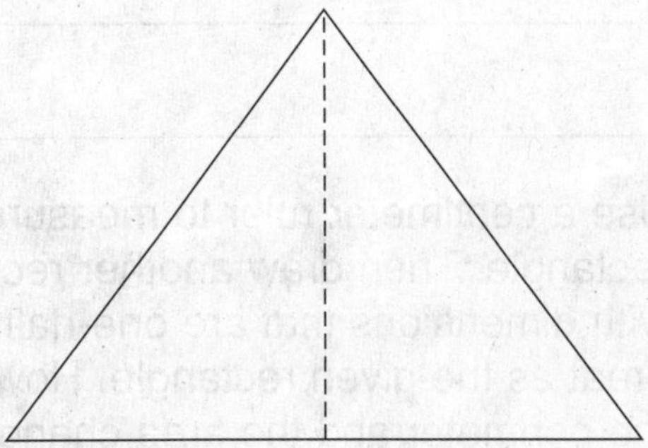

____

____

3. Nina wants to make a smaller version of a painting she saw in a museum. The museum painting was a square with each side measuring 6.4 feet. If Nina makes her copy half the size of the original painting, how much space will it cover on her wall?

____

4. How many feet of wood will Nina need to make a frame for her painting from Exercise 3?

____

Name ______________________ Date ____________ Class ____________

LESSON **10-4**

Practice C

Comparing Perimeter and Area

Write how the perimeter and the area of the figure change when its dimensions change.

1.

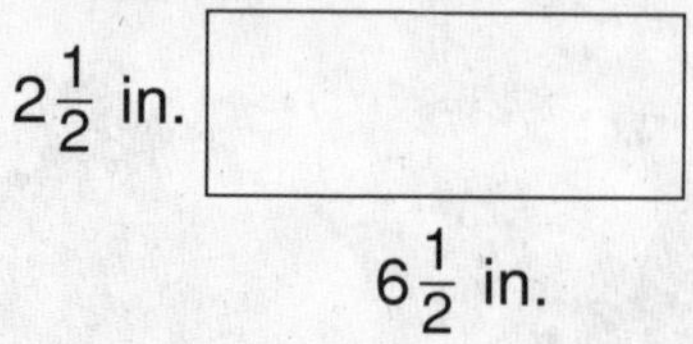

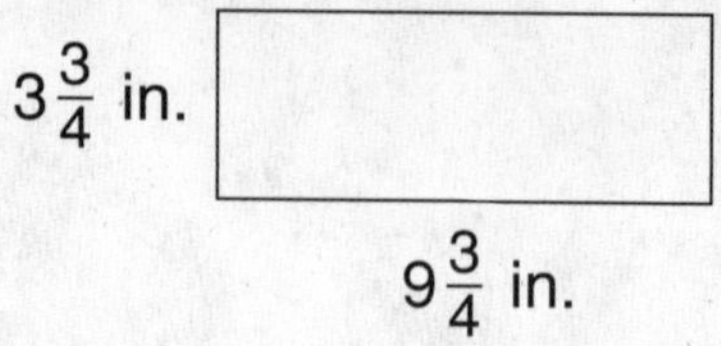

$P =$ ____________ $P =$ ____________

$A =$ ____________ $A =$ ____________

____

____

2. Use a centimeter ruler to measure the rectangle. Then draw another rectangle with dimensions that are one-half as great as the given rectangle. How do the perimeter and the area change when the dimensions change?

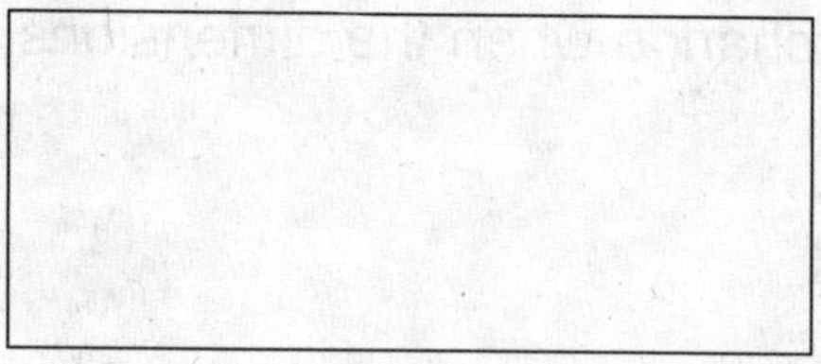

____

____

3. The area of a rectangular photo is 80 square inches. After reducing its dimensions, the area of the photo is 5 square inches. By how much were the dimensions of the original photo reduced?

____

4. The side lengths of the original and reduced photo in Exercise 3 are whole numbers. If the length of the original photo is 10 inches, what is the perimeter of the reduced photo?

____

Name ______________________ Date __________ Class __________

LESSON **10-4**

Reteach

Comparing Perimeter and Area

When the dimensions of a figure change, the perimeter and area of the figure change.

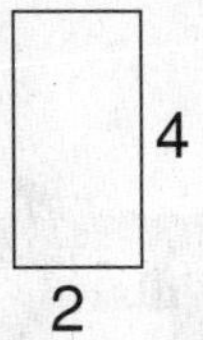

$P = 2\ell + 2w$ $\qquad$ $A = \ell w$

$= (2 \cdot 4) + (2 \cdot 2)$ $\qquad$ $= 4 \cdot 2$

$= 8 + 4$ $\qquad$ $= 8$ square units

$= 12$ units

The dimensions are $\ell = 4$ and $w = 2$.

If each dimension of the rectangle is divided by 2, the perimeter and area change.

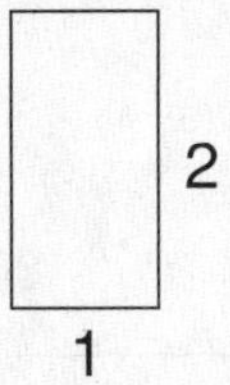

$P = 2\ell + 2w$ $\qquad$ $A = \ell w$

$= (2 \cdot 2) + (2 \cdot 1)$ $\qquad$ $= 2 \cdot 1$

$= 4 + 2$ $\qquad$ $= 2$ square units

$= 6$ units

The new dimensions are $\ell = 2$ and $w = 1$.

When the dimensions of a rectangle are divided by 2, the perimeter is divided by 2 and the area is divided by 4.

Write how the perimeter and area change when the dimensions change.

1.

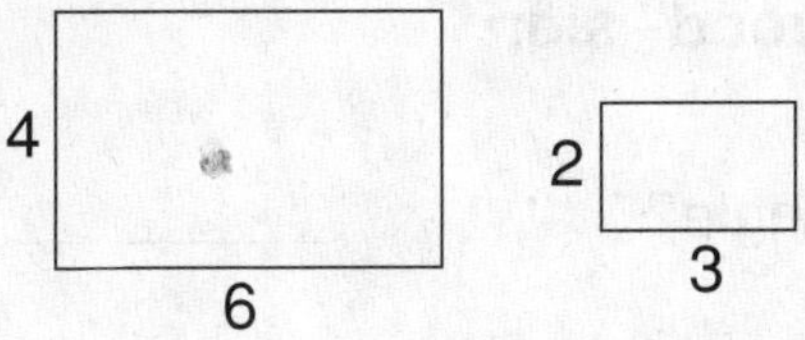

____

____

2.

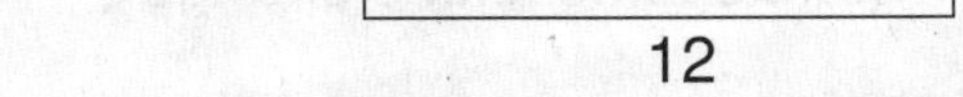

____

____

Name ______________________ Date __________ Class __________

LESSON **10-4**

Challenge

Hooray for Hollywood!

The “Hollywood” sign in Los Angeles, California, is one of the most recognized landmarks of the United States. Each of the sign’s letters covers a rectangular space 9 meters wide and 12 meters tall.

Built in 1923, the original sign said “Hollywoodland.”

Find these measurements of the “Hollywood” sign.

1. What is the perimeter of each letter’s rectangle? __________
2. What is the area of each letter’s rectangle? __________
3. What is the perimeter of the rectangle formed by the entire sign, including only the letters? __________
4. What is the area the rectangle formed by the entire sign, including only the letters? __________

Reduce the dimensions of each letter in the “Hollywood” sign by one-half.

5. What is the perimeter of each reduced letter’s rectangle? __________
6. What is the area of each reduced letter’s rectangle? __________
7. What is the perimeter of the rectangle formed by the entire reduced sign, including only the letters? __________
8. What is the area of the rectangle formed by the entire reduced sign, including only the letters? __________
9. How does the perimeter of each letter change when its dimensions change?

____

10. How does the area of the entire sign change when its dimensions change?

____

____

Name ______________________ Date __________ Class __________

LESSON 10-4

Problem Solving

Comparing Perimeter and Area

Write the correct answer.

1. Fiona's school photograph is 6 inches long and 5 inches wide. If she orders a triple enlargement how would this affect the area of the photo? How would the enlargement affect the frame she would need for the photo?

____

____

____

____

____

2. The Whitman's kitchen is 8 feet long and 6 feet wide. They are planning on renovating the kitchen to have more space. If they double just the width, how will it affect the area of the room? If they double just the length? If they double both measurements?

____

____

____

____

____

Circle the letter of the correct answer.

3. Kent saw a table in a magazine that was 3 feet wide and 4 feet long. If he wants to make a similar version of the table with an area 4 times larger, what dimensions should he use?

A 4 ft wide and 5 ft long

B 6 ft wide and 8 ft long

C 9 ft wide and 12 ft long

D 12 ft wide and 16 ft long

4. The triangular sail on Shakeera's boat is 8 meters wide and 10 meters tall. She wants to make a model of the boat that is $\frac{1}{20}$ of its actual size. How much canvas will Shakeera use for the model boat's sail?

F 10 m^2 of canvas

G 1 m^2 of canvas

H 0.1 m^2 of canvas

J 0.01 m^2 of canvas

5. A triangle is 6.4 cm long and 8.2 cm tall. If you triple its dimensions, what would be the area of the enlarged triangle?

A 78.72 cm^2 **C** 236.16 cm^2

B 157.44 cm^2 **D** 472.32 cm^2

6. The dimensions of a regular pentagon are doubled. The perimeter of the enlarged pentagon is 25 yards. What was the length of each side of the original pentagon?

F 2.5 yards **H** 5 yards

G 12 yards **J** 16.25 yards

Name ______________________ Date ____________ Class ____________

LESSON **10-4**

Reading Strategies

Compare and Contrast

Mark is trying to decide which of two different pens to buy for his dog.

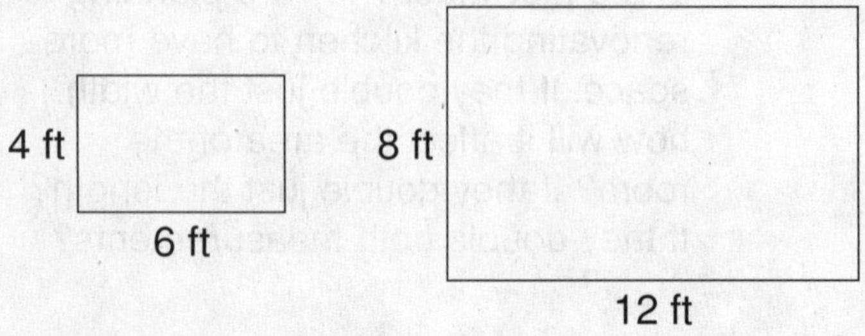

Answer the following questions.

1. Compare the shapes of the two pens. How are they alike?

____

2. Compare the lengths and widths of the two pens. How are they different?

____

____

____

3. What is the perimeter of the smaller pen? Of the larger pen?

____

4. How do the perimeters of the two pens compare in length?

____

____

5. What is the area of the smaller pen? Of the larger pen?

____

6. How do the areas of the two pens compare in size?

____

Name ______________________ Date ____________ Class ____________

LESSON **10-4**

Puzzles, Twisters & Teasers

Truth or Dare?

There are three correct answers for each problem. Circle the incorrect statement. The letters next to the incorrect statements will provide the solution to the riddle.

1. **B** If you double the dimensions, the new perimeter will be 120 cm.
   F The perimeter of the figure is 60 cm.
   R If you double the dimensions, the new area will be 432 cm^2.
   T If you halve the dimensions, the new perimeter will be 30 cm.

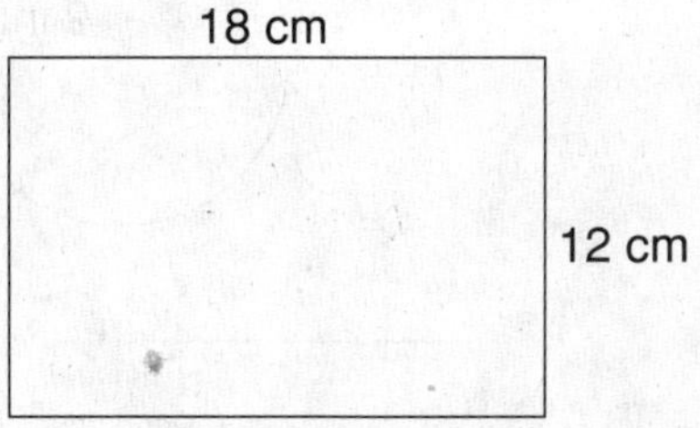

2. **A** If you triple the dimensions, the new perimeter will be 324 cm.
   C If you divide the dimensions in half, the new perimeter will be 18 cm.
   I If you triple the dimensions, the new area will be 9 times as large.
   S The area of the figure is 54 cm^2.

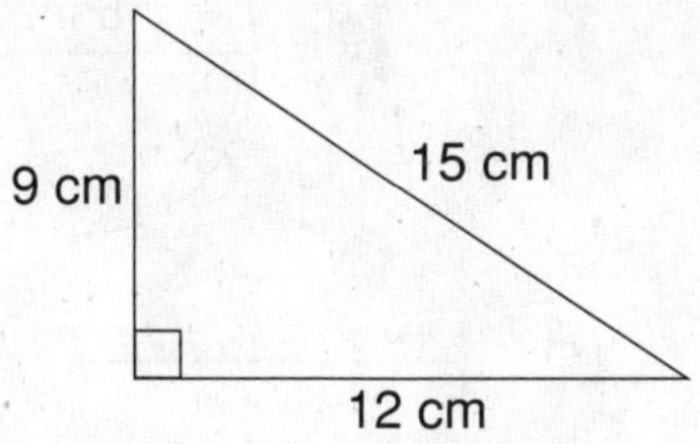

3. **L** The area of the figure is 128 ft^2.
   W The perimeter of the figure is 52 ft.
   M If you halve the dimensions, the area will be 32 ft^2.
   A If you double the dimensions, the new perimeter will be four times as large.

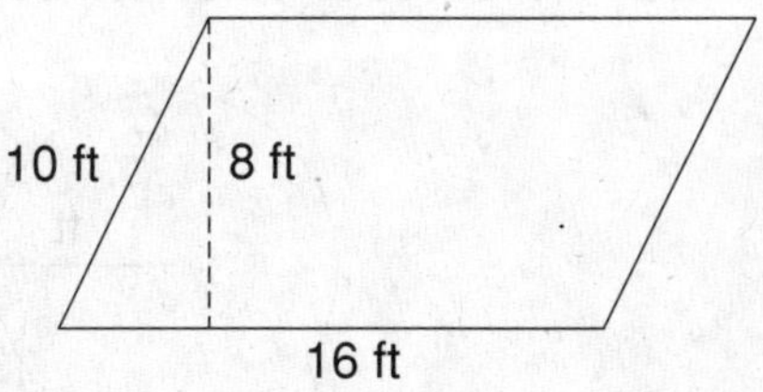

4. **W** If you multiply the dimensions by 4, the new perimeter will be 288 cm.
   S The area of the figure is 157.5 cm^2.
   E If you divide the dimensions by 3, the new area will be 35 cm^2.
   G The perimeter of the figure is 72 cm.

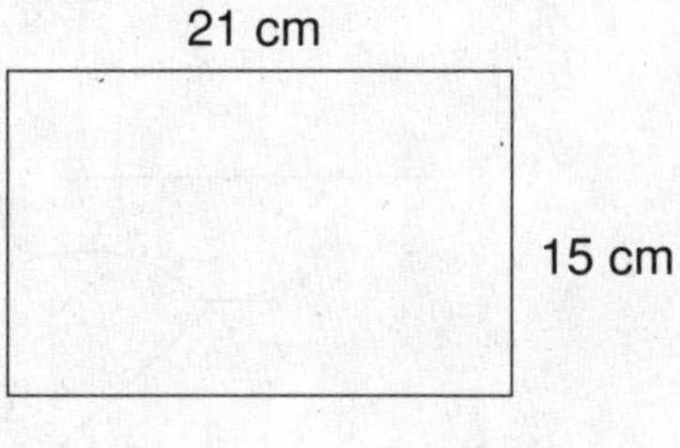

5. **P** The area of the figure is 120 ft^2.
   H The perimeter of the figure is 50 ft.
   C If you multiply the dimensions by 5, the new perimeter will be 5 times as large.
   T If you multiply the dimensions by 3, the new area will be 6 times as large.

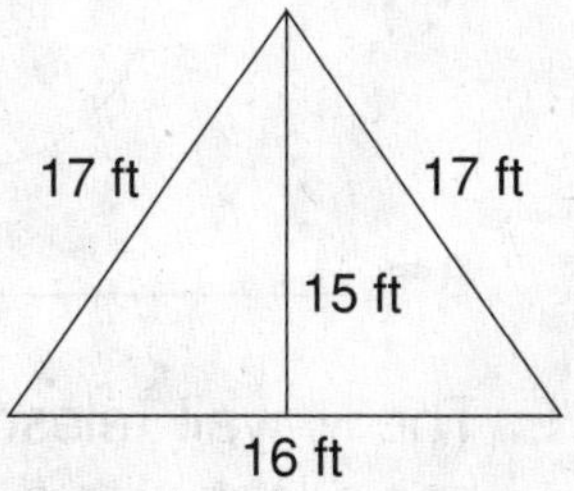

What is higher than a house and seems smaller than a mouse?

___		___	___	___	___
3		4	5	3	1

Name ______________________ Date __________ Class __________

LESSON **10-5**

Practice A
Area of Circles

Estimate the area of each circle.

1.

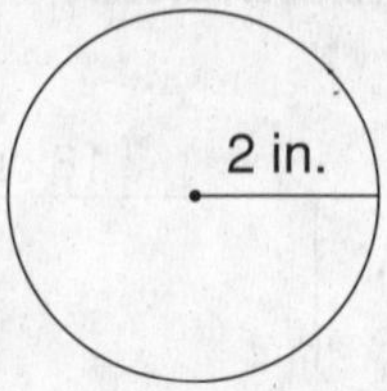

$A \approx$ __________

2.

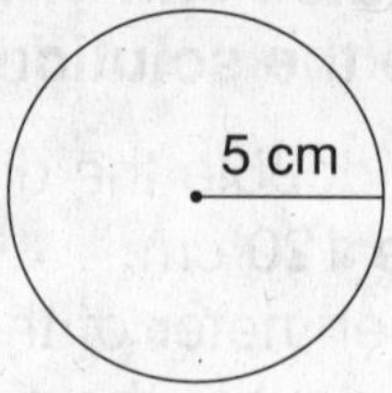

$A \approx$ __________

3.

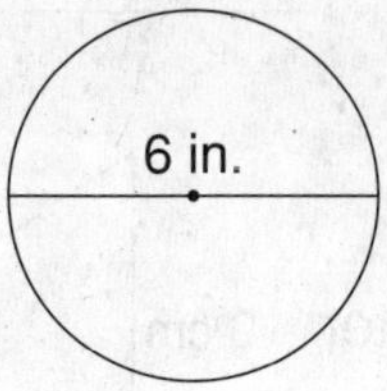

$A \approx$ __________

4.

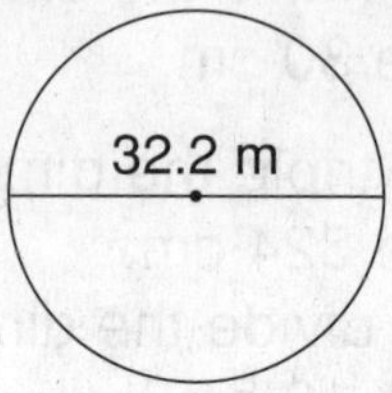

$A \approx$ __________

Find the area of each circle. Use $\frac{22}{7}$ for pi.

5.

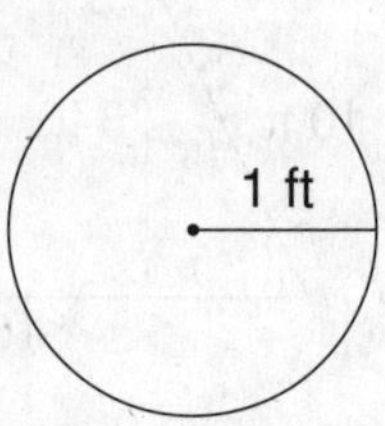

$A \approx$ __________

6.

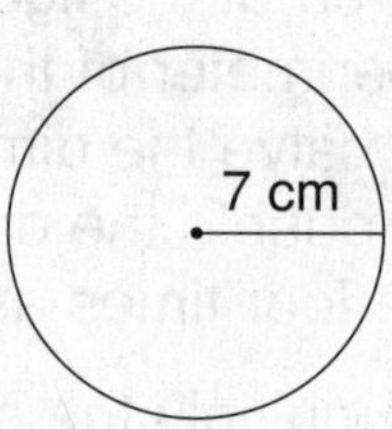

$A \approx$ __________

7.

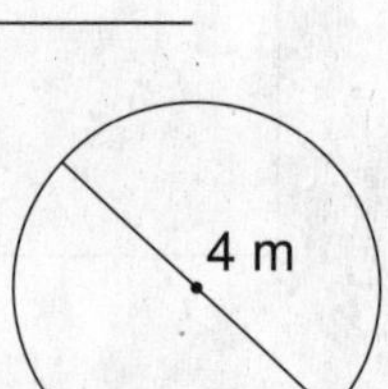

$A \approx$ __________

8.

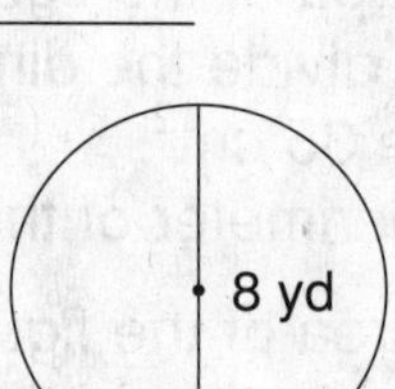

$A \approx$ __________

9. The Mayall Telescope at Kitt Peak National Observatory in Arizona has a lens with a 2-meter radius. This glass lens is covered with a thin sheet of reflective aluminum. How many square meters of glass does the aluminum cover? Use 3.14 for pi.

____

Name ______________________________ Date ____________ Class ____________

LESSON **10-5**

Practice B
Area of Circles

Estimate the area of each circle.

1.

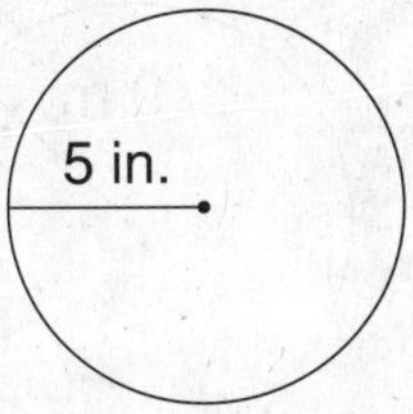

$A \approx$ ______________

2.

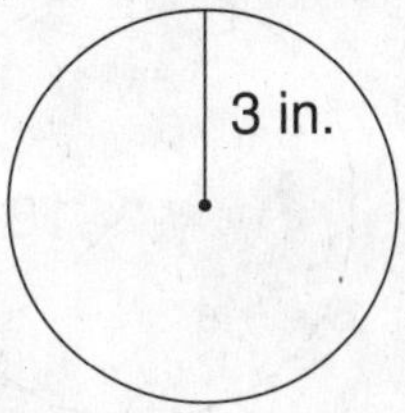

$A \approx$ ______________

3.

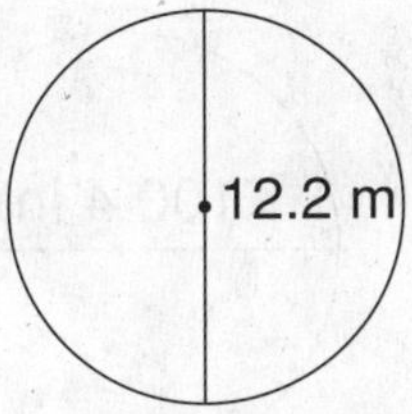

$A \approx$ ______________

4.

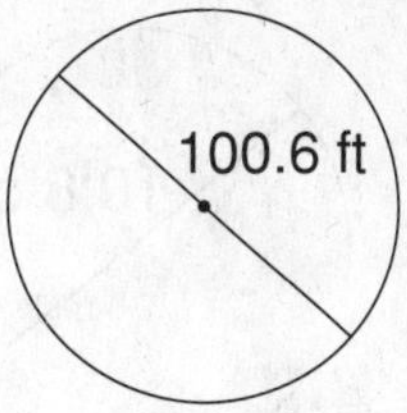

$A \approx$ ______________

Find the area of each circle. Use $\frac{22}{7}$ for *pi*.

5.

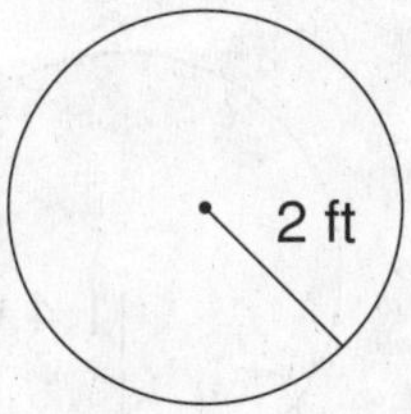

$A \approx$ ______________

6.

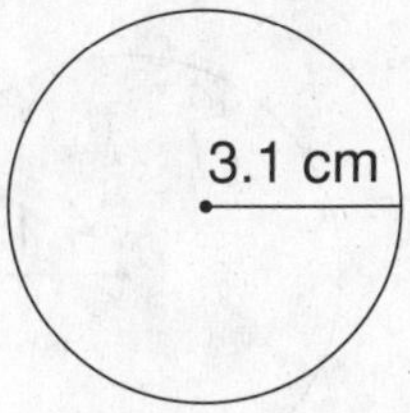

$A \approx$ ______________

7.

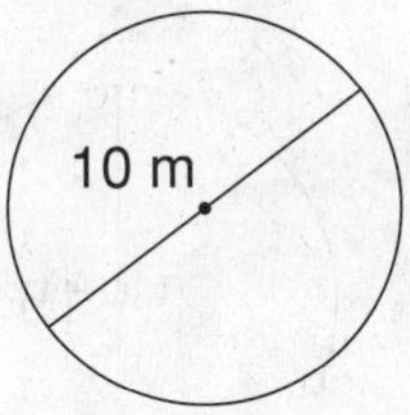

$A \approx$ ______________

8.

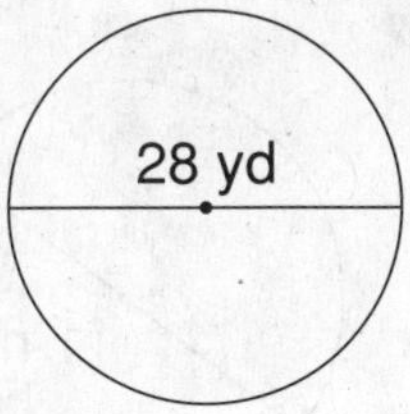

$A \approx$ ______________

9. Stonehenge, a circle of large carved stones in England, was built more than 1,000 years ago. The circle of stones has a diameter of 108 feet. About how many square feet of land does Stonehenge cover? Use $\frac{22}{7}$ for *pi*.

____

Name ______________________ Date ____________ Class ____________

LESSON 10-5

Practice C

Area of Circles

Estimate the area of each circle.

1.

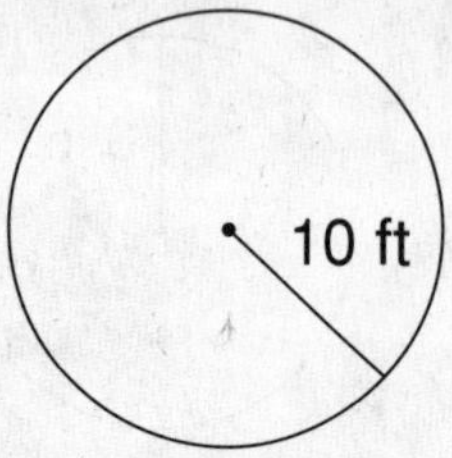

$A \approx$ ____________

2.

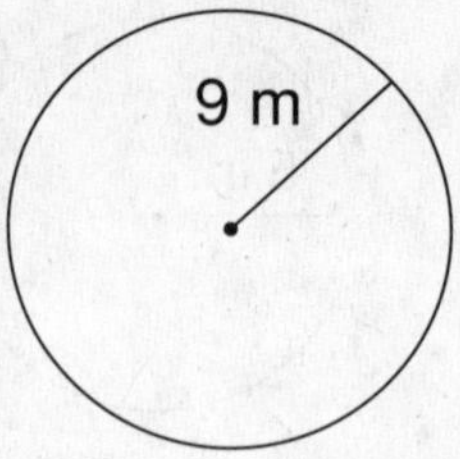

$A \approx$ ____________

3.

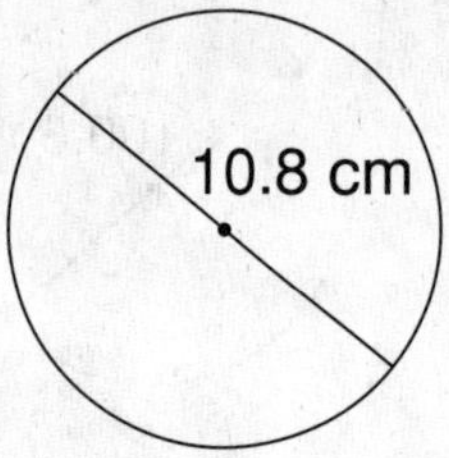

$A \approx$ ____________

4.

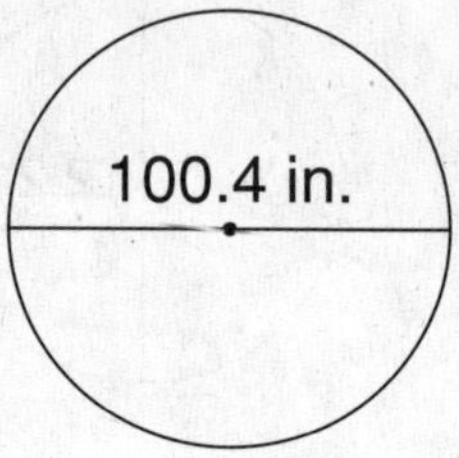

$A \approx$ ____________

Find the area of each circle. Use $\frac{22}{7}$ for *pi*.

5.

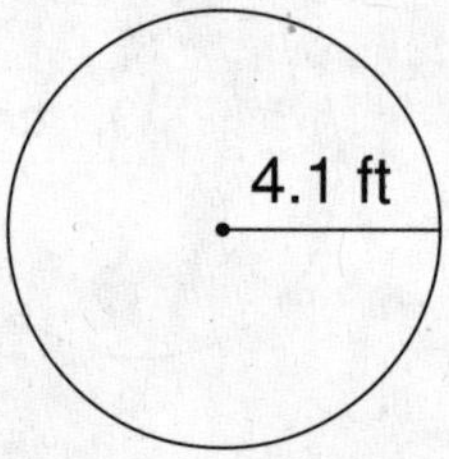

$A \approx$ ____________

6.

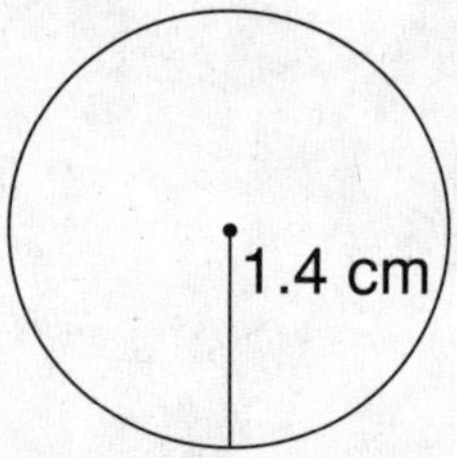

$A \approx$ ____________

7.

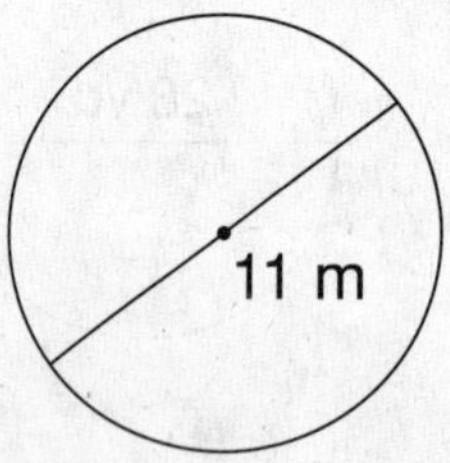

$A \approx$ ____________

8.

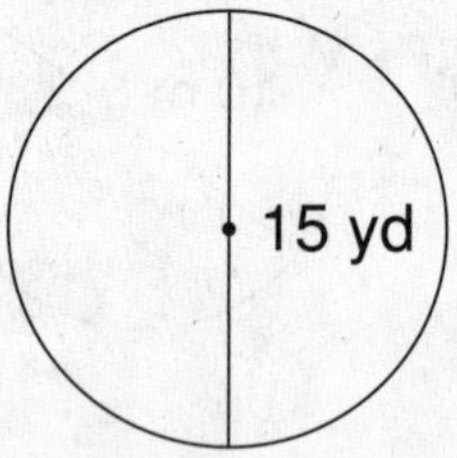

$A \approx$ ____________

9. The diameter of a sandwich cookie is 1.75 inches. To the nearest tenth, what is the combined area of the two wafer parts of the sandwich cookie? Use 3.14 for *pi*.

____

Name ______________________ Date ____________ Class ____________

LESSON **10-5**

Reteach

Area of Circles

To estimate the area A of a circle, use the formula $A = \pi r^2$ and 3 for π.

$A = \pi r^2$
$\approx 3 \cdot 2^2$
$\approx 3 \cdot 4$
$\approx$ 12 square units.

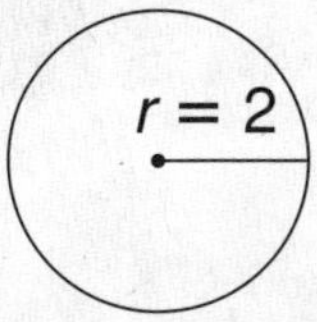

The area is about 12 square units.

Estimate the area of each circle. Use 3 for π.

1.

____

2.

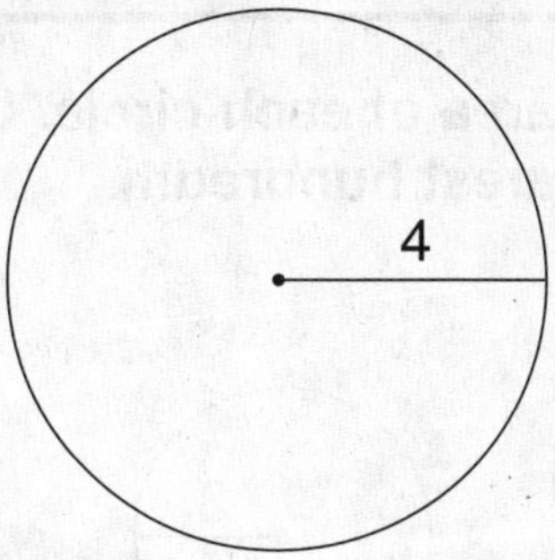

____

3.

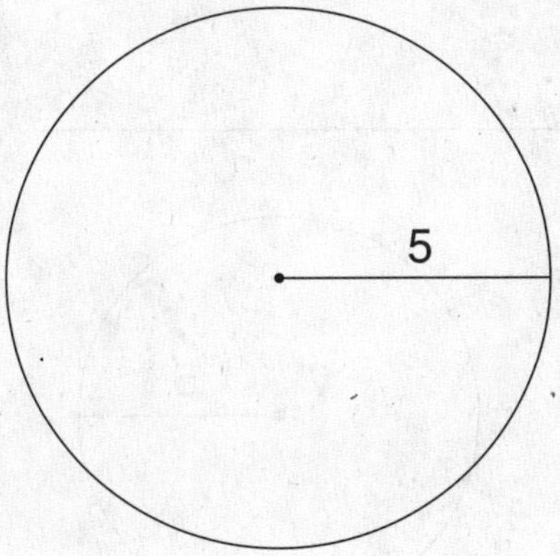

____

4.

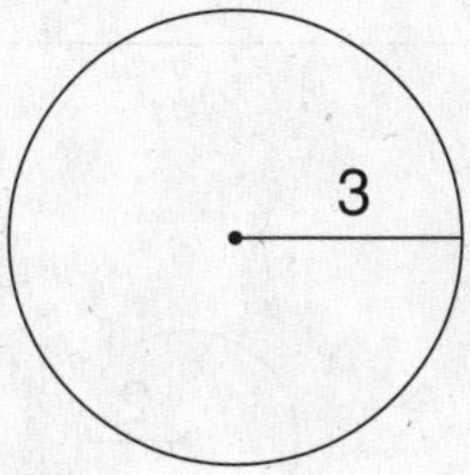

____

Name ______________________ Date ____________ Class ____________

LESSON **10-5**

Reteach

Area of Circles (continued)

To find the area A of a circle, use the formula $A = \pi r^2$ and $\frac{22}{7}$ for π.

$A = \pi r^2$

$\approx \frac{22}{7} \cdot 4^2$

$\approx \frac{22}{7} \cdot 16$

$\approx \frac{352}{7}$

≈ 50.29 square units

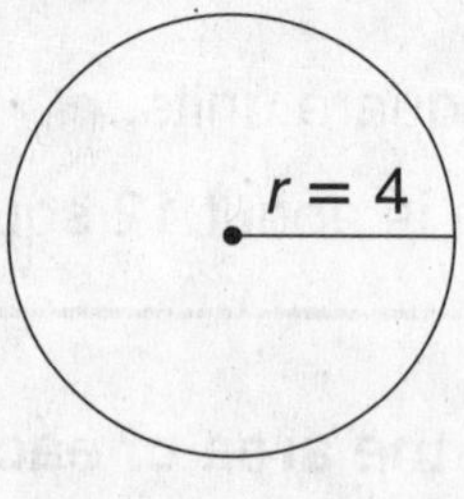

The area is about 50.29 square units.

Find the area of each circle. Use $\frac{22}{7}$ for π. Round each answer to the nearest hundredth.

5.

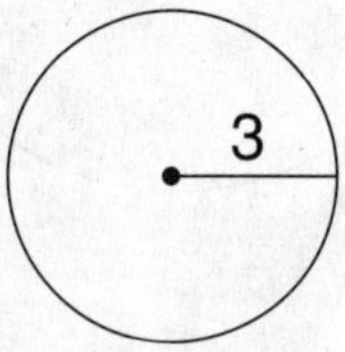

____

6.

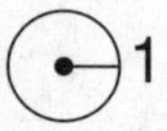

____

7.

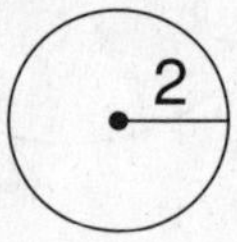

____

8.

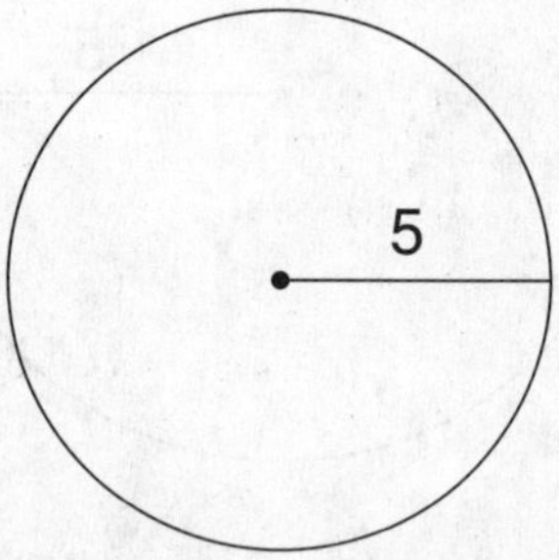

____

Name ______________________ Date ____________ Class ____________

LESSON 10-5

Challenge

Pies Are Squared

Have you ever heard the saying, "You can't fit a square peg in a round hole"? Well, sometimes you can fit a round pie in a square dish—if you squeeze really hard and don't mind messy areas.

Find the area of each pie and pie dish. Use 3.14 for π, and round your measurements to the nearest hundredth. Then imagine you could squish the pies. Which square pan would each round pie best fit in? Draw a line from the pie to the appropriate pie dish.

1.

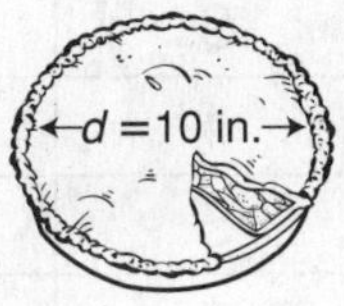

Area = ____________

A.

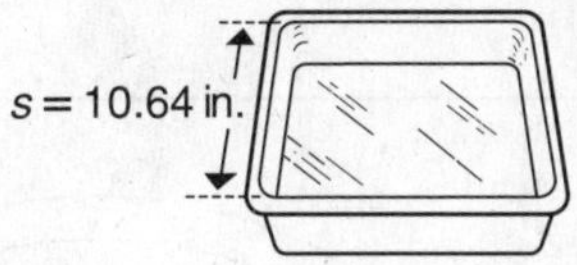

Area = ____________

2.

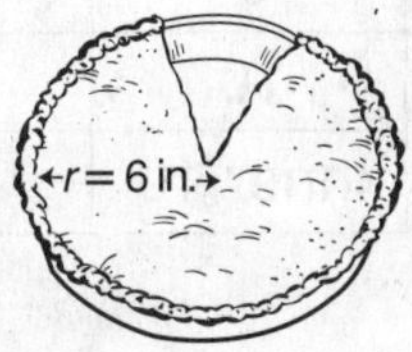

Area = ____________

B.

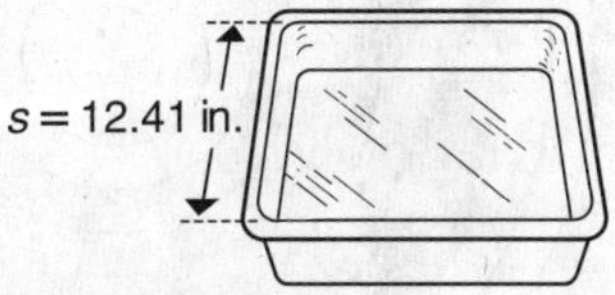

Area = ____________

3.

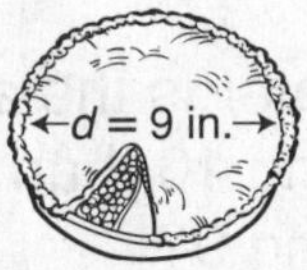

Area = ____________

C.

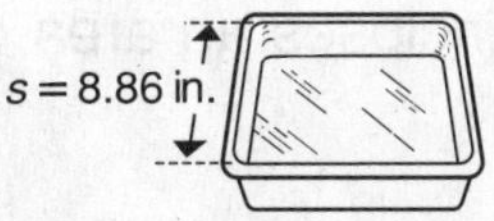

Area = ____________

4.

Area = ____________

D.

Area = ____________

Name ______________________ Date __________ Class __________

LESSON **10-5**

Problem Solving

Area of Circles

Use the table to answer each question. Use 3.14 for *pi*.

1. Which ring is the largest? What area does it enclose?

____

2. What is the area of the center circle, or the inner 10 scoring ring, on the target?

____

3. What area does Ring 5 enclose?

____

Official Archery Target Ring Diameters

Scoring Ring	Diameter (cm)
1	80
2	72
3	64
4	56
5	48
6	40
7	32
8	24
9	16
10	8
Inner 10	4

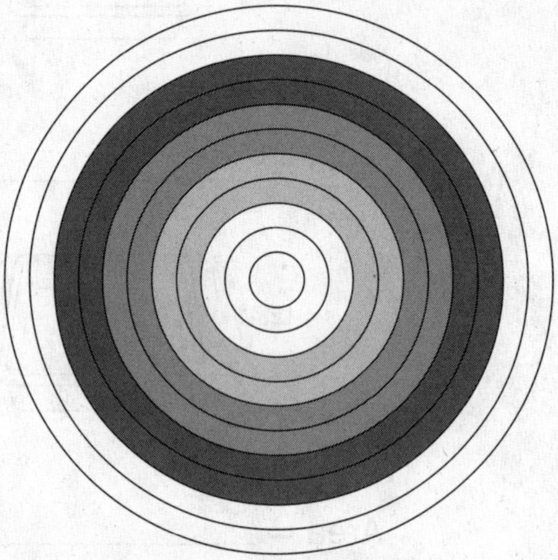

Circle the letter of the correct answer.

4. Which ring encloses an area of 4069.44 cm^2?

A Ring 2

B Ring 3

C Ring 6

D Ring 8

5. How much greater is the area enclosed by Ring 10 than the area enclosed by Ring 9?

F 50.24 cm^2

G 150.72 cm^2

H 200.96 cm^2

J 251.2 cm^2

6. What is the area enclosed by Ring 6?

A 5,024 cm^2

B 1,600 cm^2

C 1,256 cm^2

D 62.8 cm^2

7. What is the area enclosed by Ring 1?

F 10 times the area of Ring 10

G 20 times the area of Ring 10

H 100 times the area of Ring 10

J 1,000 times the area of Ring 10

Name ______________________ Date __________ Class __________

LESSON 10-5

Reading Strategies

Focus on Vocabulary

The **area** of a circle is equal to *pi* times the radius squared.

Pi is represented by the symbol π. Use $\frac{22}{7}$ or 3.14 for *pi*.

The **formula** for finding the area of a circle is $A = \pi \mathbf{r}^2$.

You can **estimate** the area of a circle by using 3 to approximate the value of *pi*.

Answer each question.

1. What is the formula for finding the area of a circle?

____

2. How do you estimate the area of a circle?

____

3. What is the symbol for *pi*?

____

4. What words can you use to tell the formula for finding the area of a circle?

____

5. A circle has a radius of 5 inches. Estimate the area.

____

6. Use 3.14 to find the area of the circle in Problem 5.

____

7. A circle has a radius of 7 centimeters. Estimate the area.

____

8. Use $\frac{22}{7}$ to find the area of the circle in Problem 7.

____

Name ______________________ Date ______________ Class ______________

LESSON 10-5

Puzzles, Twisters & Teasers

You're All Wet

Find the area of each circle below. Use 3.14 for *pi*. To solve the riddle, match the letter of the area to the problem number.

Area (square units)	Letter
63.59	O
65.39	I
94.99	M
99.49	N
30.18	T
31.08	S
132.67	T
123.67	R
19.63	O
131.04	E
113.04	B

1.

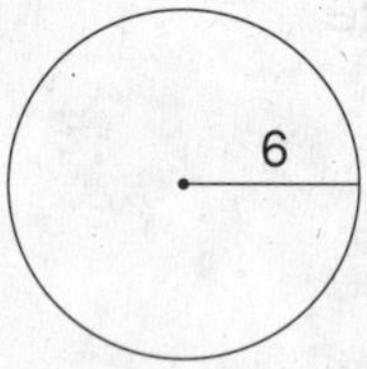

____

2.

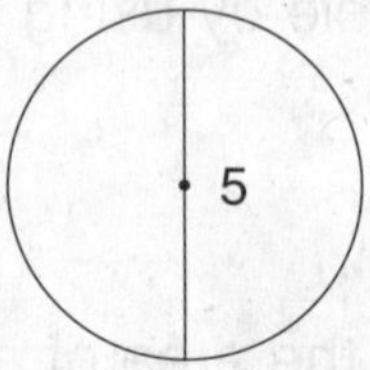

____

3.

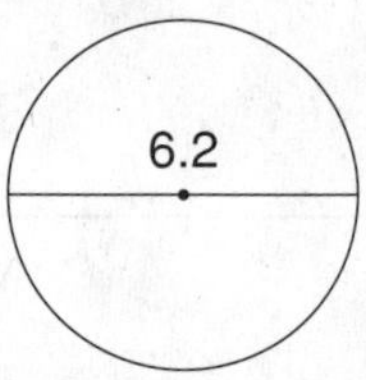

____

4.

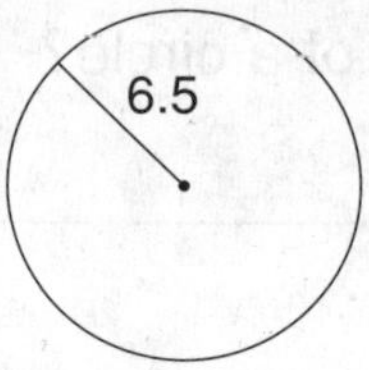

____

5.

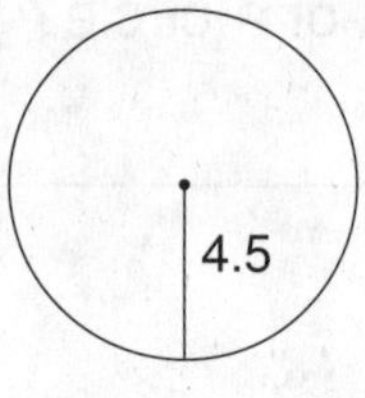

____

6.

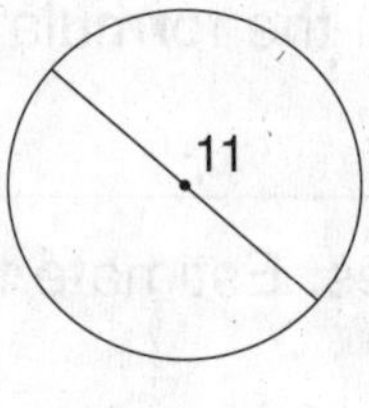

____

Where is the ocean the deepest?

ON THE ___ ___ ___ ___ ___ ___

1. 2. 3. 4. 5. 6.

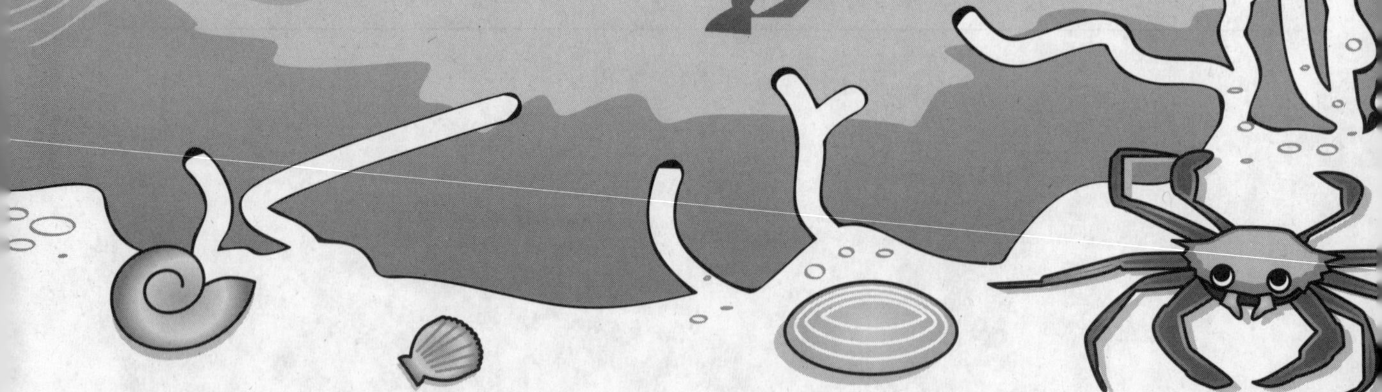

Name ______________________ Date ____________ Class ____________

LESSON **10-6**

Practice A

Three-Dimensional Figures

Use the polyhedron to answer each question.

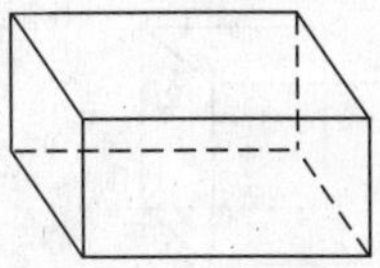

1. How many faces does it have? ________

2. How many edges does it have? ________

3. How many vertices does it have? ________

4. What is the shape of its bases?

____

5. What kind of polyhedron is it?

____

Tell whether each figure is a polyhedron and name the three-dimensional figure.

6.

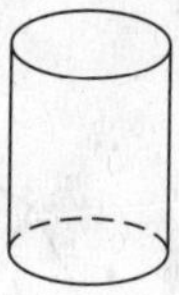

____

____

7.

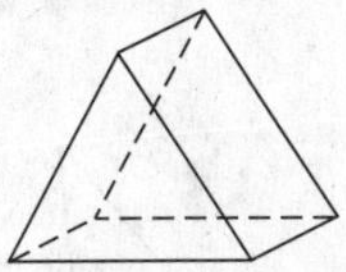

____

____

8.

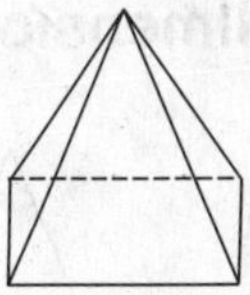

____

____

9. LeJuan bought a book for his sister's birthday. What shape box should LeJuan put the book in before wrapping it?

____

10. Look around your classroom. Find 5 three-dimensional figures and name what kind each is.

____

____

____

____

Name ______________________ Date ____________ Class ____________

LESSON **10-6**

Practice B

Three-Dimensional Figures

Identify the number of faces, edges, and vertices in each three-dimensional figure.

1.

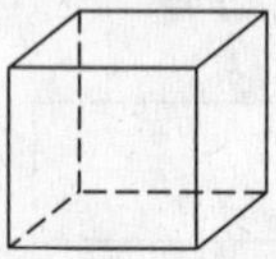

2.

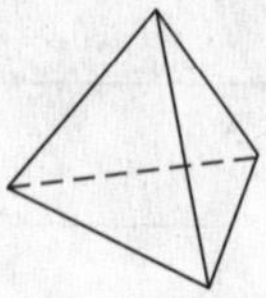

3.

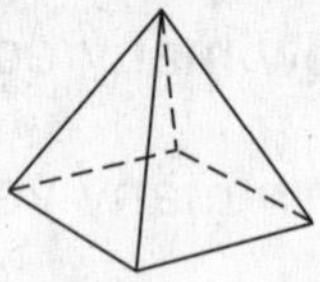

Tell whether each figure is a polyhedron and name the three-dimensional figure.

4.

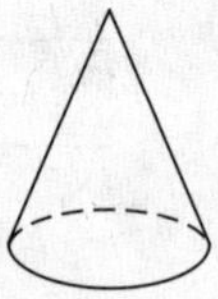

5.

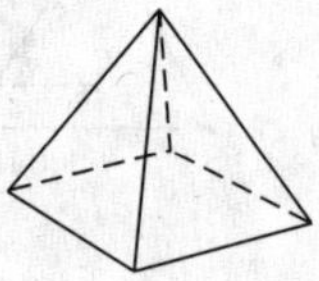

6.

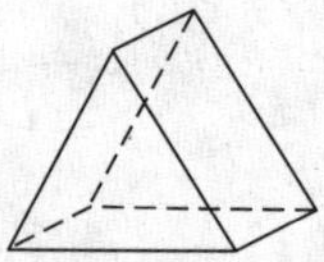

7. Kelly wants to make a box in the shape of a cube. How many pieces of wood does she need? In what shape should she cut them? Explain.

8. Kwan made a sculpture in the shape of a polyhedron. It only has one base that is a triangle. What three-dimensional figure is her sculpture? Explain your reasoning.

Name ______________________ Date ____________ Class ____________

LESSON **10-6**

Practice C

Three-Dimensional Figures

Tell whether each figure is a polyhedron and name the three-dimensional figure. Then identify the number of faces, edges, and vertices in each three-dimensional figure.

1.

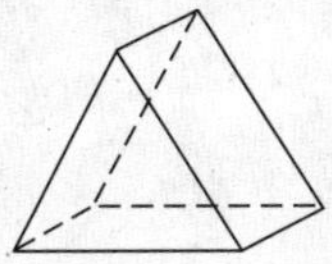

2.

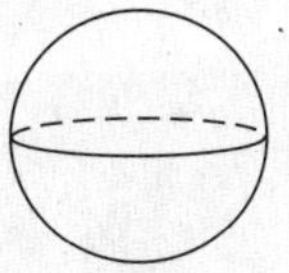

3.

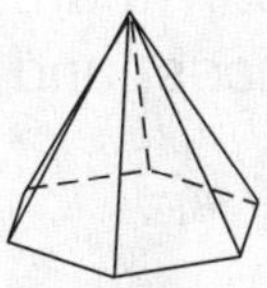

4.

5.

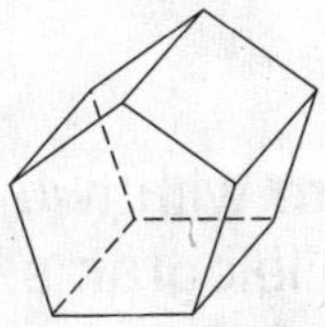

6.

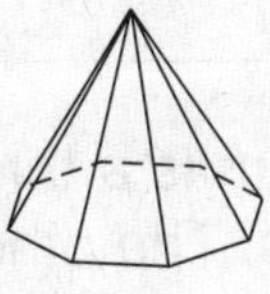

7. A construction company is building two hotels that will be the same height, and have bases that are the same size. One will be a rectangular prism and the other will be a rectangular pyramid. Which building will require more gallons of paint to completely cover it? Why?

Name ____________________ Date __________ Class __________

LESSON **10-6**

Reteach

Three-Dimensional Figures

A polyhedron is a three-dimensional figure. A polyhedron has flat surfaces, called faces. Faces that share a side form an edge. The point at which three or more edges meet is called a vertex.

This polyhedron has 4 faces, 4 vertices, and 6 edges.

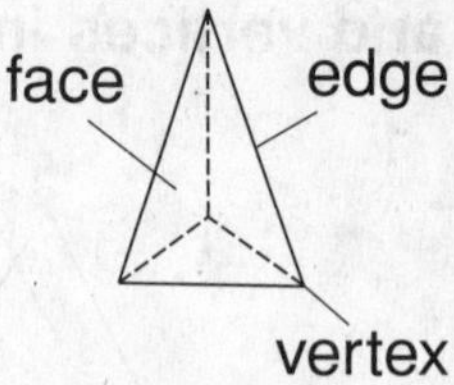

Identify the number of faces, edges, and vertices on each three-dimensional figure.

1.

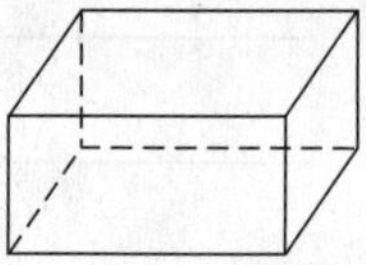

____

2.

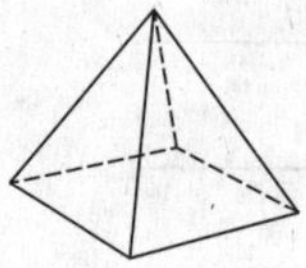

____

A prism is a three-dimensional figure with two congruent parallel bases. The other faces are all parallelograms.

A pyramid has one base. The other faces are triangles.

A cylinder has two circular parallel bases joined by a curved surface.

A cone has one circular base and a curved surface that comes to a point.

Pyramids and prisms are polyhedrons because each face is a polygon. Cylinders and cones are not polyhedrons because they have curved surfaces.

Tell whether each figure is a polyhedron and name each three-dimensional figure.

3.

____

4.

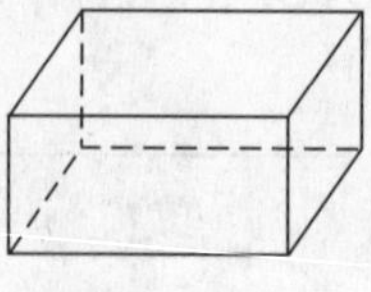

____

Name ______________________ Date ____________ Class ____________

LESSON **10-6**

Challenge
Polyhedron Patterns

Complete these charts to discover the polyhedron patterns.

	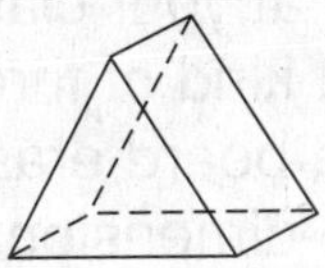 **Triangular Prism**	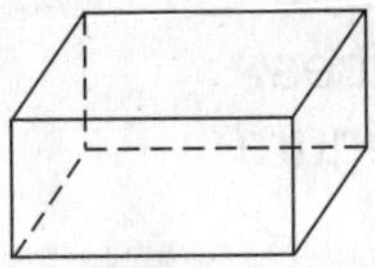 **Rectangular Prism**	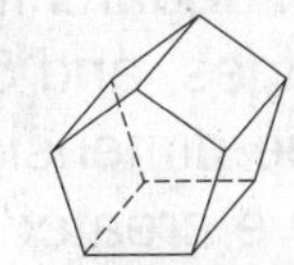 **Pentagonal Prism**	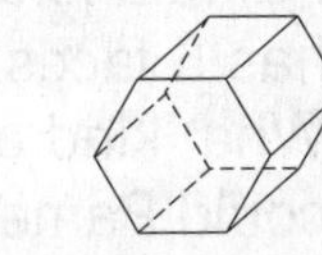 **Hexagonal Prism**
Base's Number of Sides				
Faces				
Vertices				
Edges				

PRISM PATTERNS: If $n =$ the number of sides on the base of a prism, what three expressions show that prism's number of faces, vertices, and edges?

____

	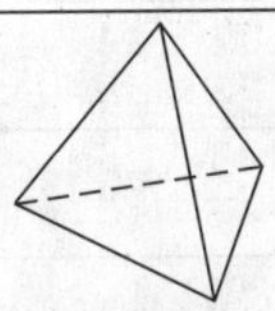 **Triangular Pyramid**	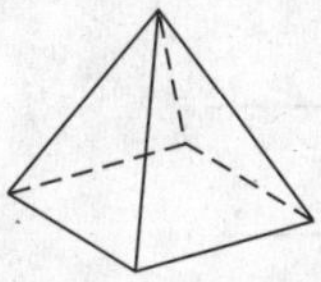 **Rectangular Pyramid**	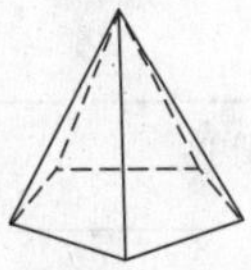 **Pentagonal Pyramid**	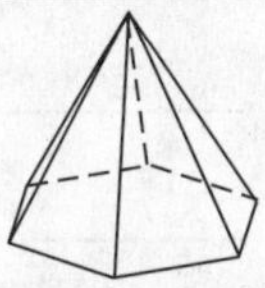 **Hexagonal Pyramid**
Base's Number of Sides				
Faces				
Vertices				
Edges				

PYRAMID PATTERNS: If $n =$ the number of sides on the base of a pyramid, what three expressions show that pyramid's number of faces, vertices, and edges?

____

Name ______________________ Date ____________ Class ____________

LESSON 10-6

Problem Solving

Three-Dimensional Figures

Write the correct answer.

1. Pamela folded an origami figure that has 5 faces, 8 edges, and 5 vertices. What kind of three-dimensional figure could Pamela have created?

____

____

2. Look at your classroom chalkboard. What kind of three-dimensional figure is the board eraser? What kind of three-dimensional figure is the chalk?

____

____

3. If you cut a cylinder in half between its two bases, what two three-dimensional figures are formed?

____

4. You have two hexagons. How many rectangles do you need to create a hexagonal prism?

____

5. All four of the faces of a paperweight are triangles. Is this enough information to classify this three-dimensional figure? Explain.

____

____

____

6. Paulo says that if you know the number of faces a pyramid has, you also know how many vertices it has. Do you agree? Explain.

____

____

____

Circle the letter of the correct answer.

7. How is a triangular prism different from a triangular pyramid?

A The prism has 2 bases.

B The pyramid has 2 bases.

C All of the prism's faces are triangles.

D The pyramid has 5 faces.

8. Which of these statements is not true about a cylinder?

F It has 2 circular bases.

G It has a curved lateral surface.

H It is a solid figure.

J It is a polyhedron.

9. A museum needs to ship a sculpture that has a curved lateral surface and one flat circular base. In what shape box should they mail the sculpture?

A cone

B cube

C cylinder

D triangular prism

10. A glass prism reflects white light as a multicolored band of light called a spectrum. The prism has 5 glass faces with 9 edges and 6 vertices. What kind of prism it it?

F cube

G cone

H triangular pyramid

J triangular prism

Name ______________________ Date ____________ Class ____________

LESSON **10-6**

Reading Strategies

Use a Graphic Organizer

This graphic organizer will help you learn about three-dimensional figures called **polyhedrons.**

Polyhedrons

Definition
A three-dimensional geometric figure that has four or more flat faces.

Prism
- Has two congruent parallel bases.
- All faces are polygons.
- Named by shape of its base.

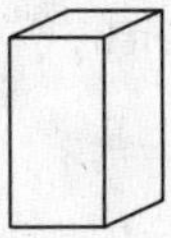

Pyramid
- Has one polygon-shaped base.
- Other faces are triangles.
- Triangular faces meet at a common vertex.
- Named by the shape of its base.

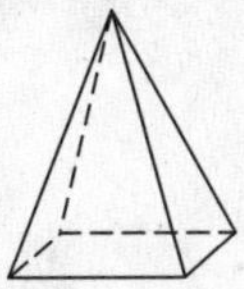

Non-examples
Cylinder and cone
- Bases are circular.
- Surface is curved.

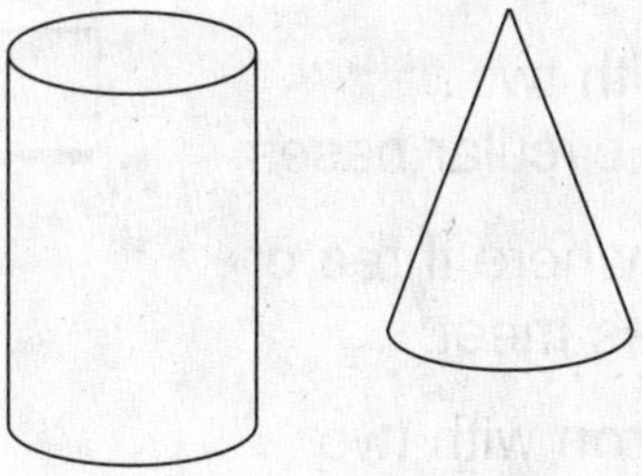

Use the graphic organizer to answer each question.

1. What are the flat surfaces of a polyhedron called? ____________

2. What shape are the faces, other than base, of a pyramid? ____________

3. What shape are the faces of a prism? ____________

4. How many bases does a pyramid have? A prism? ____________

Name ______________________ Date ____________ Class ____________

LESSON **10-6**

Puzzles, Twisters & Teasers

Crossword

Across

2. Name of the flat surface of a polyhedron.
5. Three-dimensional object with flat surfaces.
7. When two faces of a three-dimensional figure share a side they form this.
9. Figure with one polygon shaped base; the other faces are triangles that come to a point.

Down

1. A prism is named for the shape of this.
3. Three-dimensional figure with a circular base and surface that comes to a point.
4. A figure with two congruent circular bases.
6. The point where three or more edges meet.
8. A polyhedron with two congruent parallel bases.

Name ______________________ Date __________ Class __________

LESSON 10-7
Practice A
Volume of Prisms

Find the volume of each rectangular prism.

1.

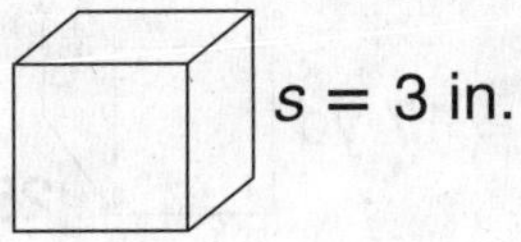

____

2.

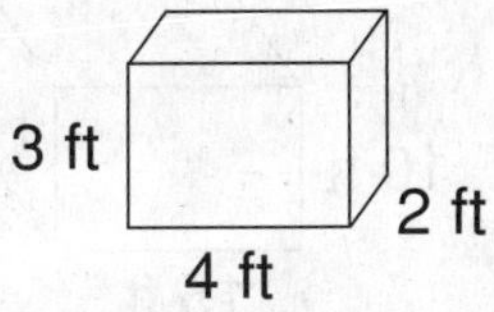

____

3.

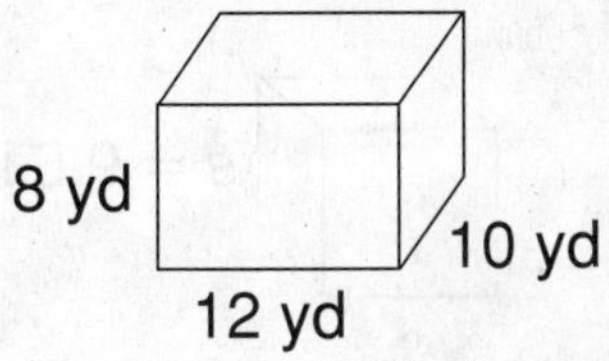

____

4.

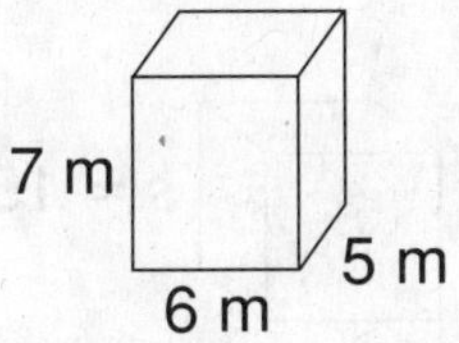

____

5.

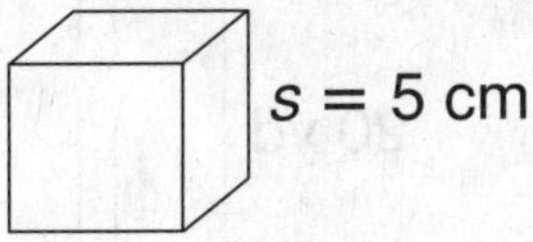

____

6.

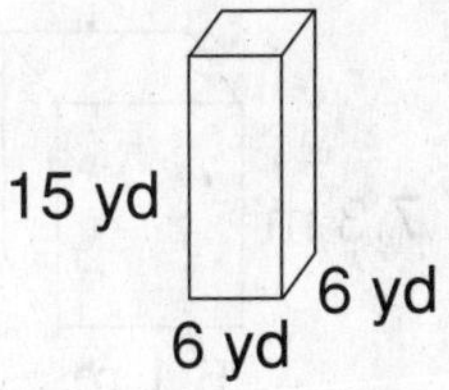

____

Find the volume of each triangular prism.

7.

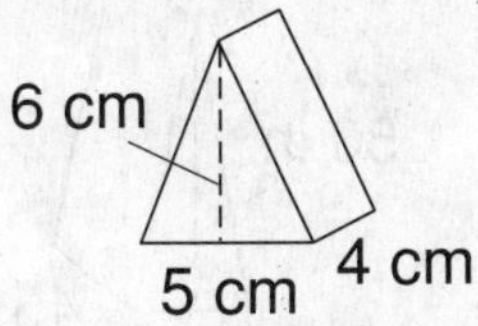

____

8.

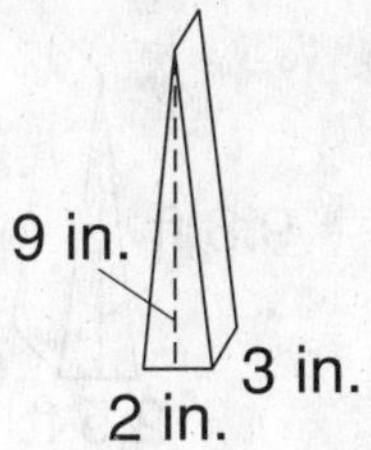

____

9.

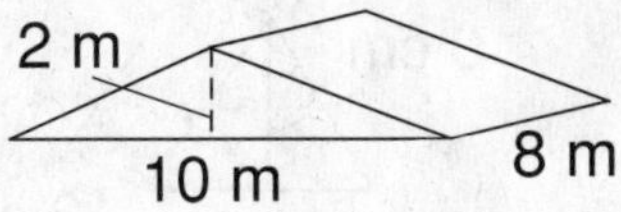

____

10. A triangular prism and a rectangular prism have the same lengths, heights, and widths. Which prism has the greater volume?

____

11. Tim made a toy chest for his little sister's square building blocks. If 6 layers of blocks can fit in the box, and each layer has 15 blocks, how many building blocks can the toy chest hold in all?

____

Name ______________________ Date ____________ Class ____________

LESSON 10-7

Practice B
Volume of Prisms

Find the volume of each rectangular prism.

1.

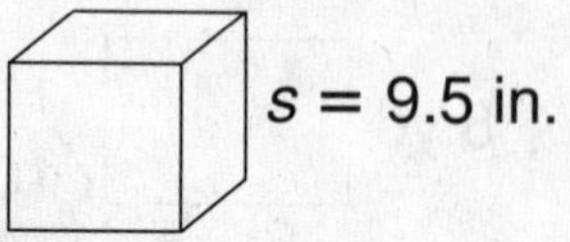

____

2.

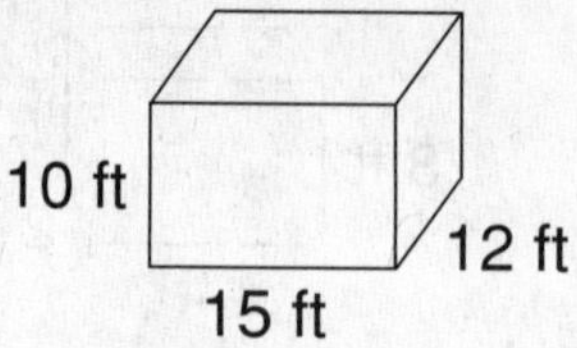

____

3.

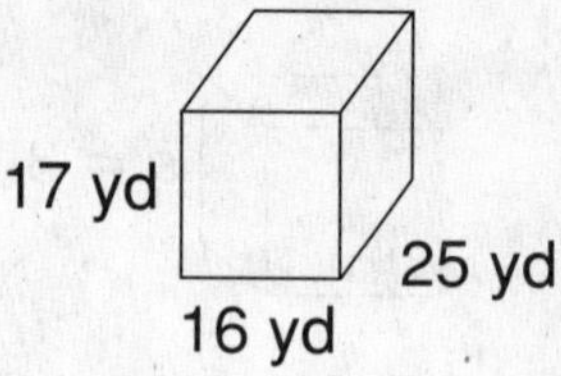

____

4.

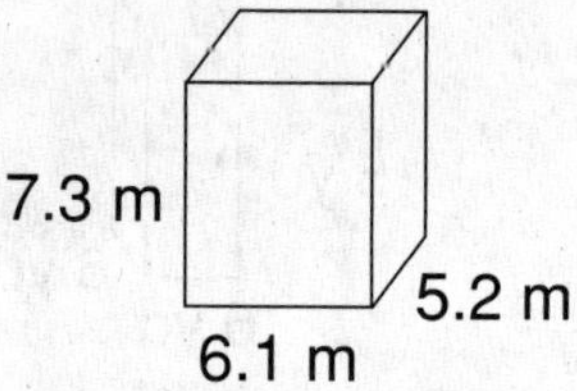

____

5.

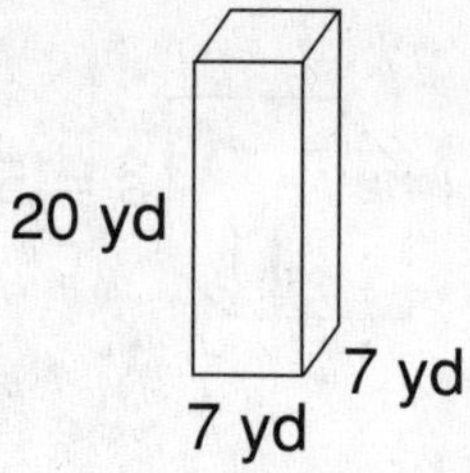

____

6.

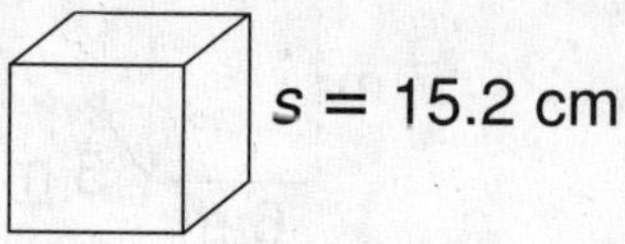

____

Find the volume of each triangular prism.

7.

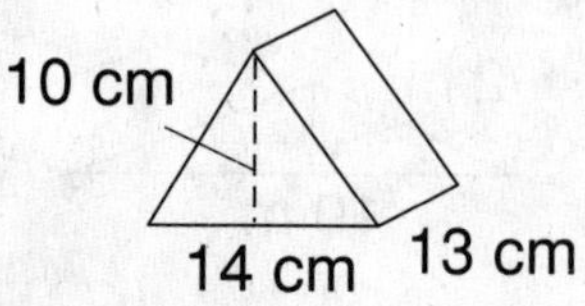

____

8.

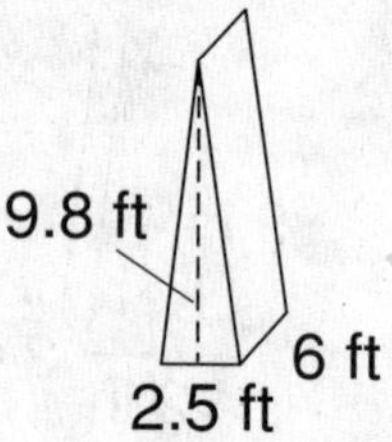

____

9.

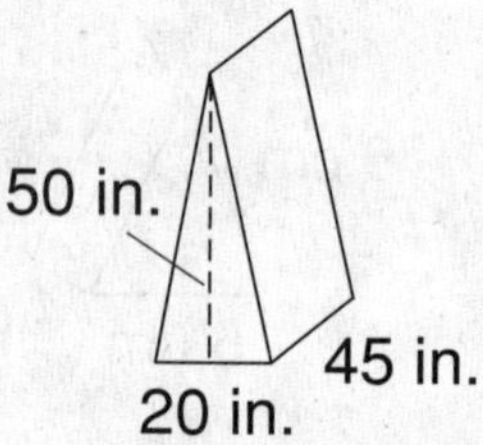

____

10. Fawn built a sandbox that is 6 feet long, 5 feet wide, and $\frac{1}{2}$ foot tall. How many cubic feet of sand does she need to fill the box?

____

11. Unfinished lumber is sold in units called board feet. A board foot is the volume of lumber contained in a board 1 inch thick, 1 foot wide, and 1 foot long. How many cubic inches of wood are in 1 board foot?

____

Name ______________________________ Date ____________ Class ____________

LESSON **10-7**

Practice C

Volume of Prisms

Find the volume of each rectangular prism.

1.

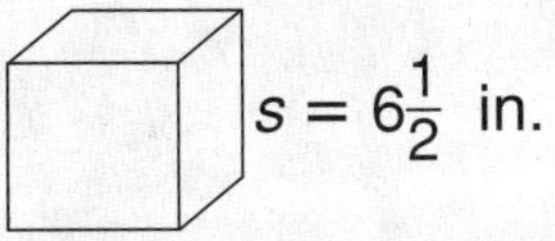

2.

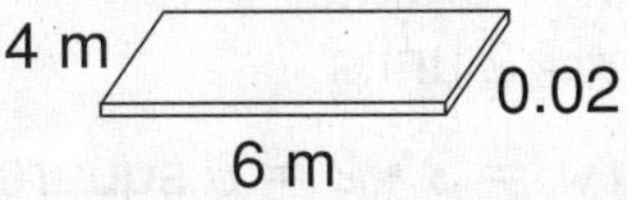

3.

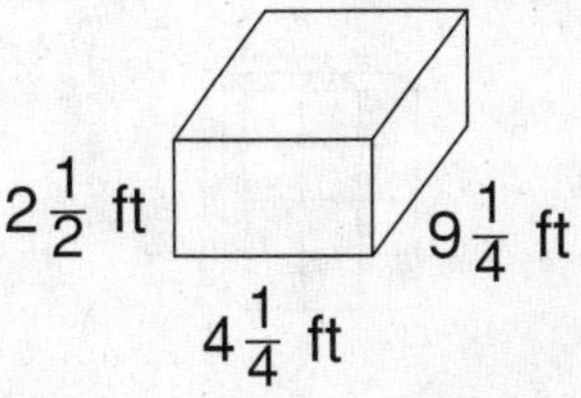

____

Find the volume of each triangular prism.

4.

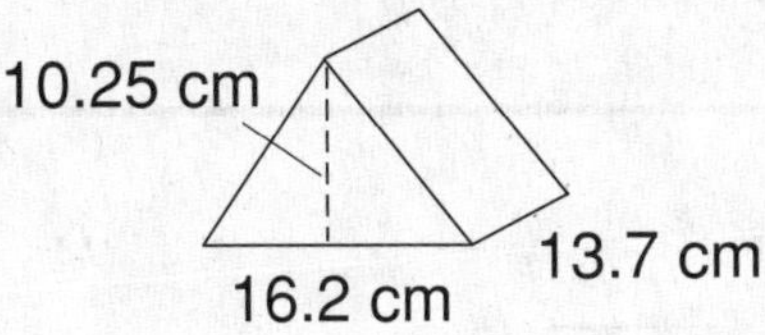

5.

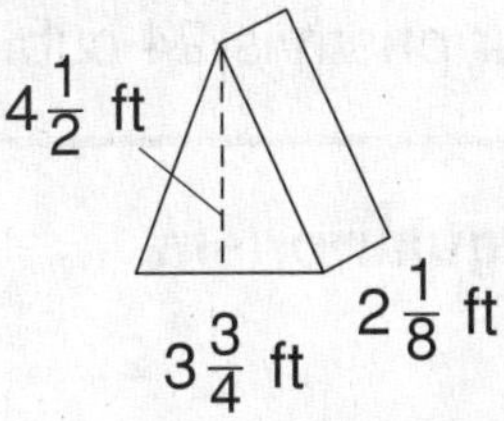

6.

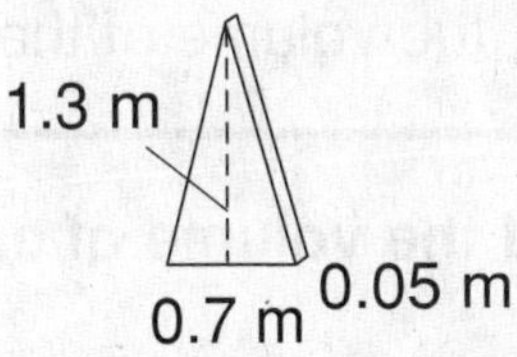

______________________ ______________________ ______________________

Find the missing measurement for each rectangular prism.

7. $\ell =$? ; $w = 34$ m; $h = 5$ m; $V = 340$ m^3 ______________

8. $\ell = 4.5$ ft; $w =$? ; $h = 6$ ft; $V = 256.5$ ft^3 ______________

9. $\ell = 14$ yd; $w = 8$ yd; $h =$? ; $V = 1{,}344$ yd^3

10. $\ell =$? ; $w = 1\frac{1}{2}$ in.; $h = 3\frac{1}{2}$ in.; $V = 52\frac{1}{2}$ in^3 ______________

11. A rectangular prism's base is 8 feet long. It is $2\frac{1}{2}$ times taller than it is long and $\frac{1}{2}$ as wide as it is tall. What is the volume of that prism? ______________

12. A rectangular box is 3 feet wide, 6 feet long, and 4 feet tall. Inside the box, Phil placed a triangular prism that is 3 feet tall, and has a base 2 feet wide with a height of 5 feet. If he places another triangular prism of the same size in the box, how much empty space will there be in the box? ______________

Name ______________________ Date ___________ Class ___________

LESSON **10-7**

Reteach

Volume of Prisms

Volume is the number of cubic units needed to fill a space. To find the volume of a rectangular prism, first find the area of the base.

length = 3 units
width = 2 units

$A = \ell w = 3 \cdot 2 = 6$ square units.

The area of the base tells you how many cubic units are in the first layer of the prism.

Next, multiply the result by the number of layers in the prism. The prism has 4 layers, so multiply 6 by 4.

$6 \cdot 4 = 24$

So, the volume of the rectangular prism is 24 cubic units.

Find the volume of each rectangular prism.

1.

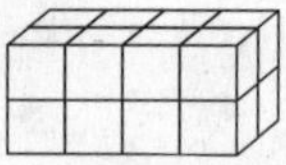

____

2.

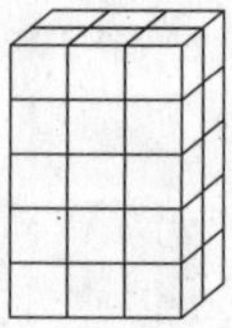

____

To find the area of a triangular prism, first find the area of the base.

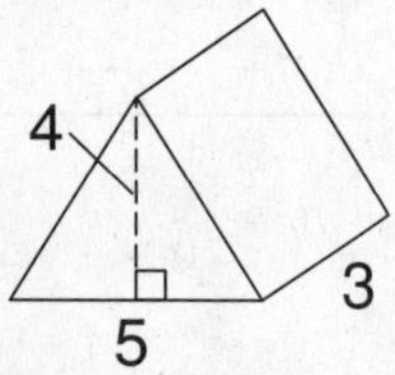

$A = \frac{1}{2}bh$

$= \frac{1}{2}(5 \cdot 4)$

$= 10$ square units

Then multiply the result by the height of the prism.

$10 \cdot 3 = 30$

The volume of the triangular prism is 30 cubic units.

Find the volume of each triangular prism.

3.

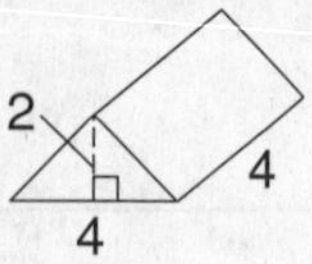

____

4.

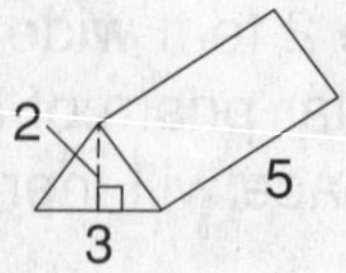

____

Name ______________________________ Date __________ Class __________

LESSON 10-7

Challenge
Fish Tank Math

Did you know that most fish grow according to the size of their tank? In fact, a freshwater fish needs about 1 gallon of water for every inch of its body. The volume of 1 gallon of water is about 231 in^3. So, a fish tank with a volume of a 700 in^3 will hold $700 \div 231 \approx 3$ gallons of water. This size tank would be the healthiest home for a 3-inch-long fish.

Find the volume of each fish tank. Then match each fish to the best tank for it.

Angelfish
$\ell = 6$ in.

Koi Goldfish
$\ell = 24$ in.

Redtail Shark
$\ell = 5$ in.

Clown Loach
$\ell = 12$ in.

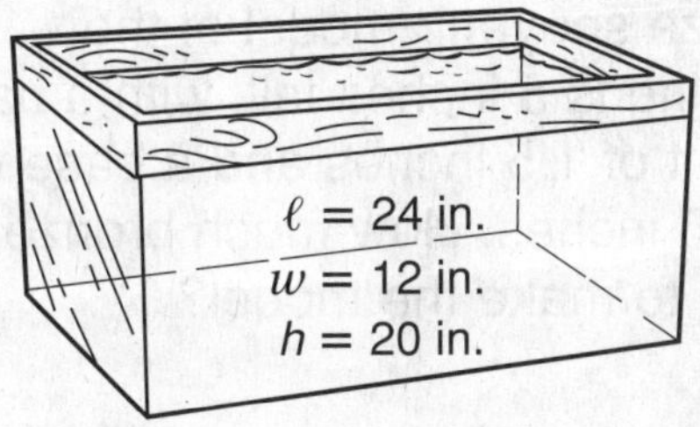

Volume of Tank: ____________

Gallons of Water: ____________

Fish: ____________

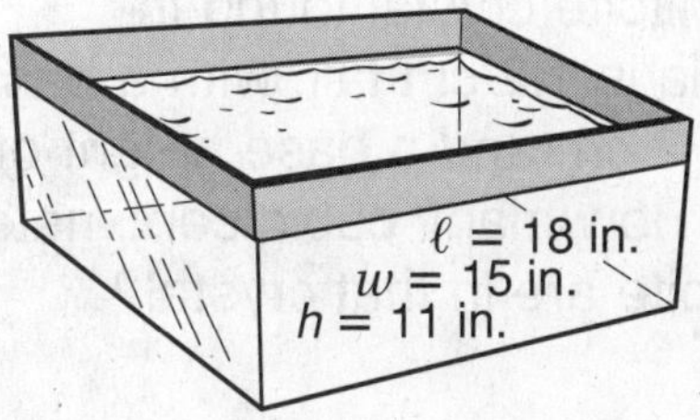

Volume of Tank: ____________

Gallons of Water: ____________

Fish: ____________

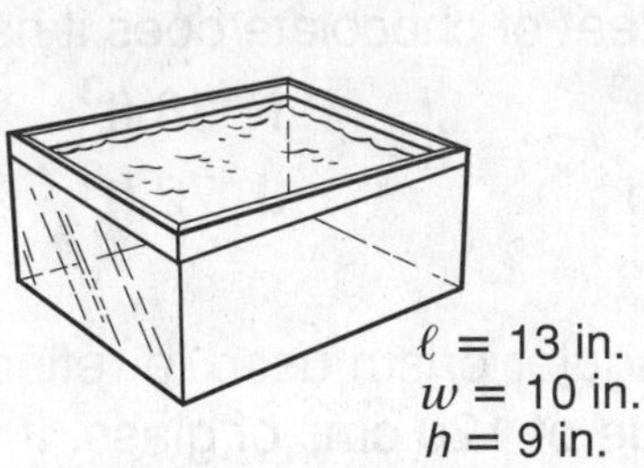

Volume of Tank: ____________

Gallons of Water: ____________

Fish: ____________

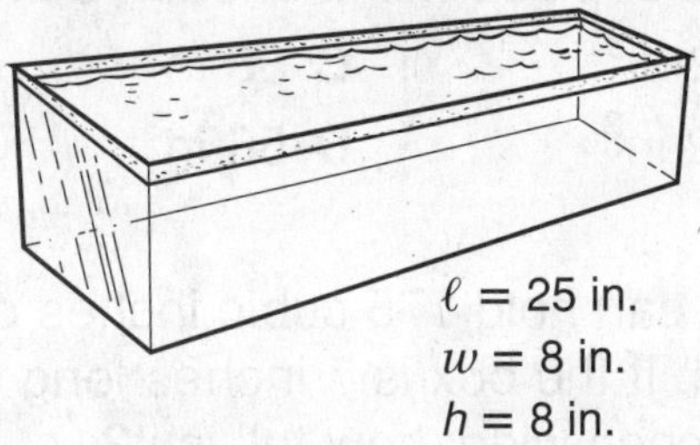

Volume of Tank: ____________

Gallons of Water: ____________

Fish: ____________

Name _______________________________ Date ___________ Class ___________

LESSON **10-7**

Problem Solving

Volume of Prisms

Write the correct answer.

1. At 726 feet tall, Hoover Dam is one of the world's largest concrete dams. In fact, it holds enough concrete to pave a two-lane highway from New York City to San Francisco! The dam is shaped like a rectangular prism with a base 1,224 feet long and 660 feet wide. About how much concrete forms Hoover Dam?

____

____

2. The Vietnam Veterans Memorial in Washington, D.C., is a 493.5-foot-long wall made of polished black granite engraved with the names of soldiers who died in the war. The wall is 0.25 feet thick and has an average height of 9 feet. About how many cubic feet of black granite was used in the Vietnam Veterans Memorial?

____

____

3. Benitoite, a triangular prism crystal, is the official state gem of California. One benitoite crystal found in California is 1.2 cm tall, with a base width of 2 cm and a base height of 1.3 cm. How many cubic centimeters of benitoite are in that crystal?

____

4. The Flatiron Building in New York City is a triangular prism. A solid bronze souvenir model of the building is 5 inches tall, with a base height of 1.5 inches and a base width of 2.5 inches. How much bronze was used to make the model?

____

Circle the letter of the correct answer.

5. Individual slices of pizza are sold in 2-inch-tall triangular prism boxes. The box base is 8 inches wide, with a 7-inch height. How many cubic inches of pizza will fit in each box?

A 112 in^3 **C** 60 in^3

B 102 in^3 **D** 56 in^3

6. The world's largest chocolate bar is a huge rectangular prism weighing more than a ton! The bar is 9 feet long, 4 feet tall, and 1 foot wide. How many cubic feet of chocolate does it have?

F 13 ft^3 **H** 36 ft^3

G 14 ft^3 **J** 72 ft^3

7. A box can hold 175 cubic inches of cereal. If the box is 7 inches long and 2.5 inches wide, how tall is it?

A 25 in.

B 10 in.

C 17.5 in.

D 9.5 in.

8. A triangular prism used to reflect light is made of 120 cm^3 of glass. If the prism is 5 centimeters tall, what is the area of each of its triangular bases?

F 24 cm

G 12 cm

H 12 cm^2

J 24 cm^2

Name ______________________ Date __________ Class __________

LESSON **10-7**

Reading Strategies
Analyze Information

The **volume** of a prism is the number of cubic units it contains.

Look at the first layer of cubic units in this rectangular prism.

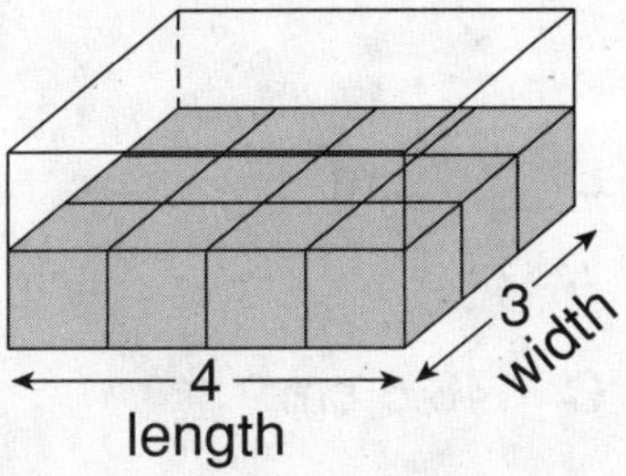

1. How many cubes long is the first layer? ______________

2. How many cubes wide is the first layer? ______________

3. How many cubes are in the first layer? ______________

Now look at the next layer of cubes.

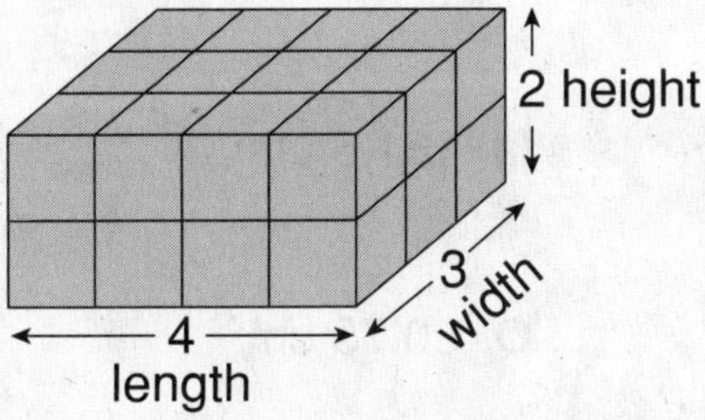

4. How many cubes long is the top layer? ______________

5. How many cubes wide is the top layer? ______________

6. How many cubes are in the top layer? ______________

7. What is the total number of cubes in both layers? ______________

8. What is the total volume of the rectangular prism? ______________

Name ______________________ Date ____________ Class ____________

LESSON **10-7**

Puzzles, Twisters & Teasers

3D Match

Find the volume of each of the prisms below. Draw a line to the correct answer. Rearrange the letters to solve the riddle.

1.

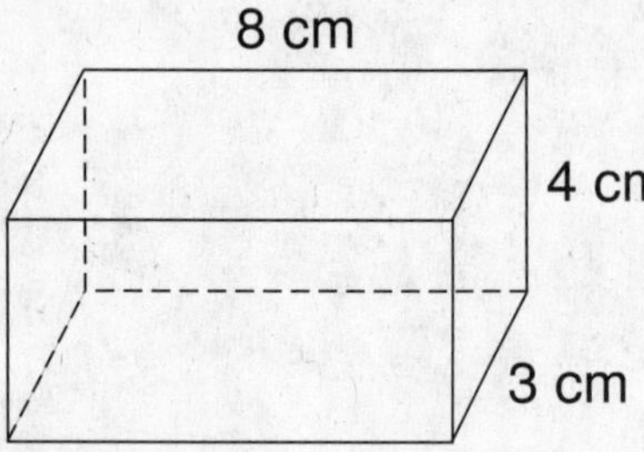

2.

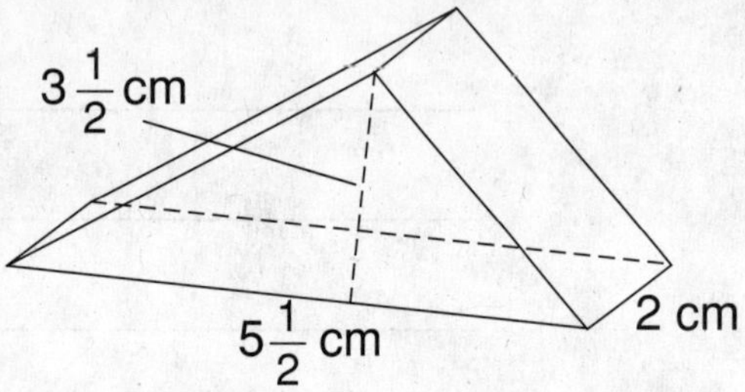

3.

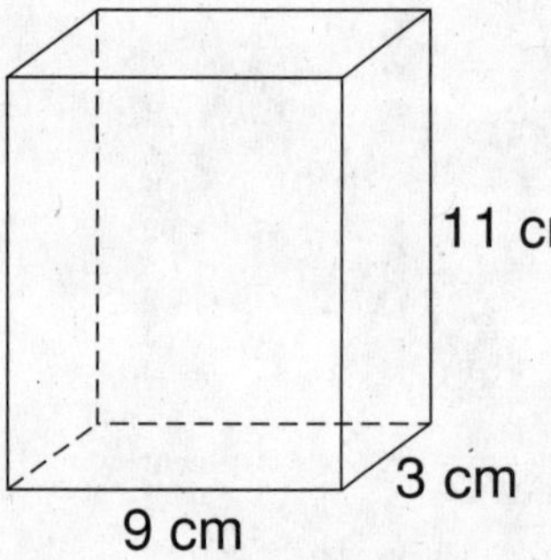

4.

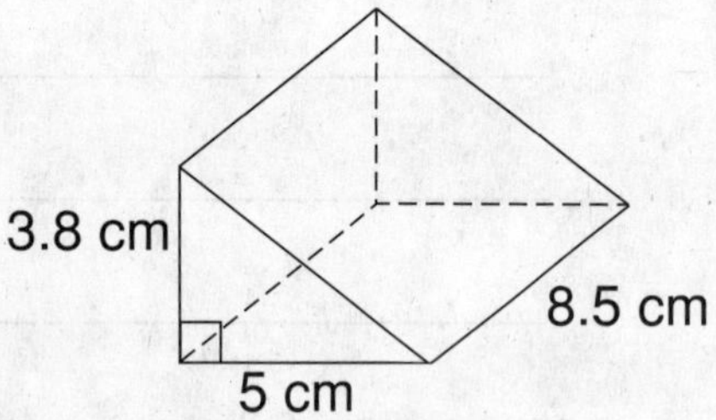

5.

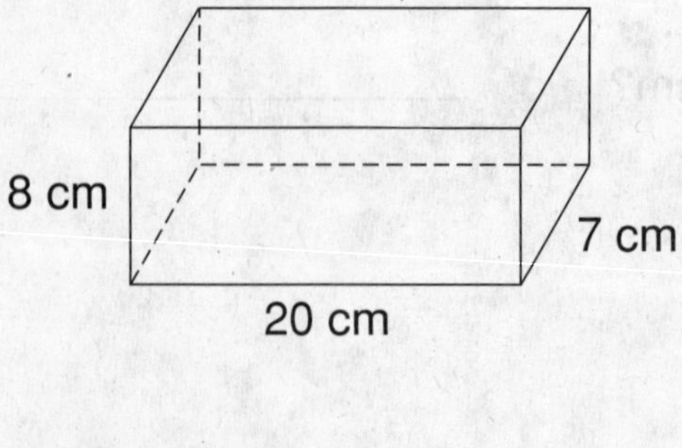

L 297 cm^3

C 148.5 cm^3

G 43 cm^3

H 96 cm^3

A $19\frac{1}{4}$ cm^3

M 161.5 cm^3

O 80.75 cm^3

S 30.25 cm^3

E 1120 cm^3

K 560 cm^3

Name ______________________ Date ____________ Class ____________

LESSON **10-8**

Practice A

Volume of Cylinders

Find the volume *V* of each cylinder to the nearest cubic unit.

1.

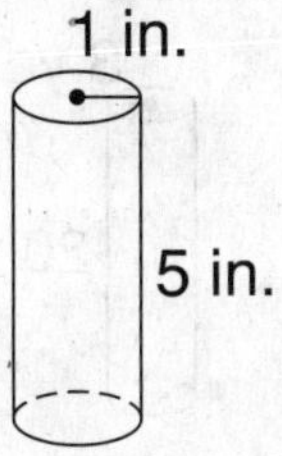

____

2.

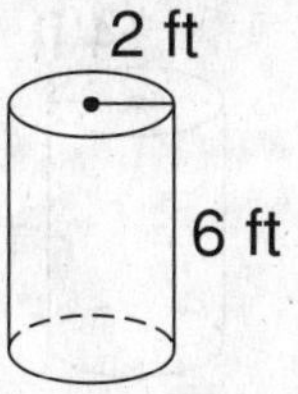

____

3.

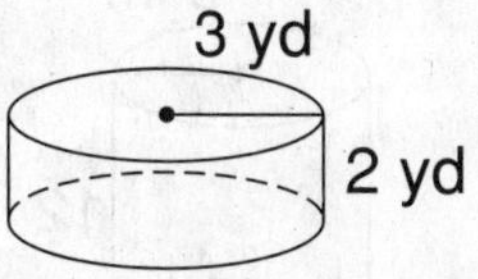

____

4.

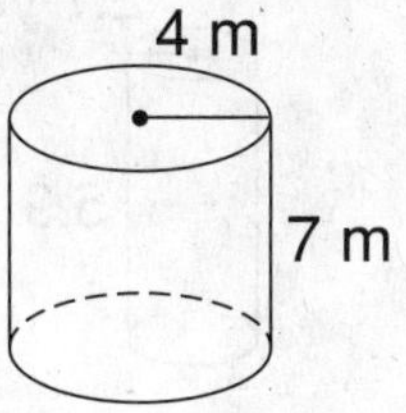

____

5.

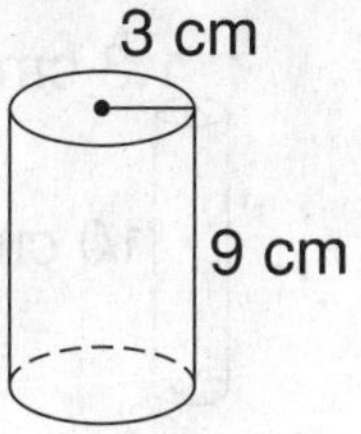

____

6.

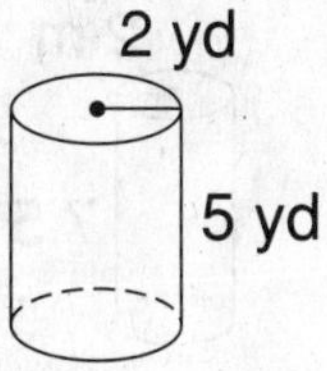

____

7.

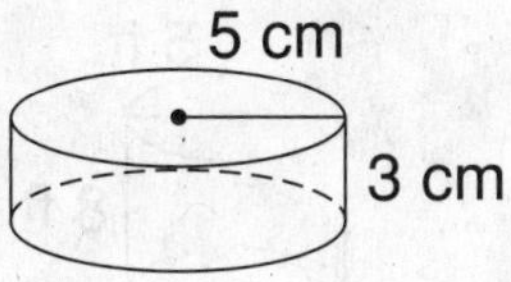

____

8.

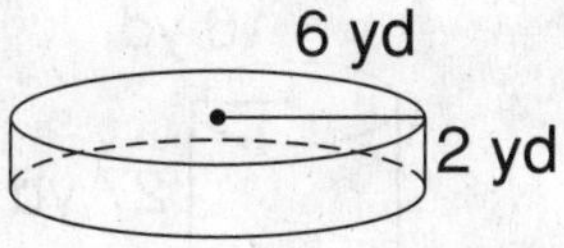

____

9.

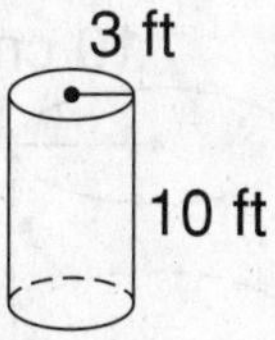

____

10. Which of the following could be the measurement for the volume of a can: 20 centimeters, 20 cubic centimeters, or 20 square centimeters?

____

11. The diameter of a tube is 4 centimeters. It is 40 centimeters tall. What is the volume of the tube?

____

Name ______________________ Date __________ Class __________

LESSON **10-8**

Practice B

Volume of Cylinders

Find the volume V of each cylinder to the nearest cubic unit.

1.

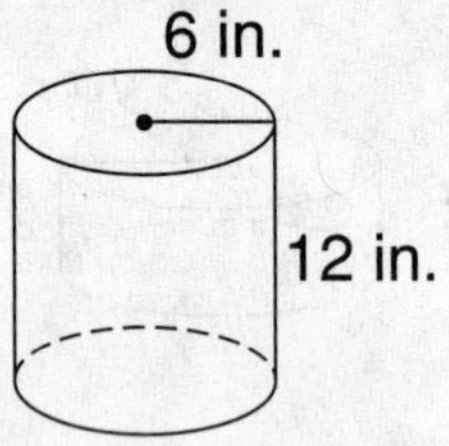

____

2.

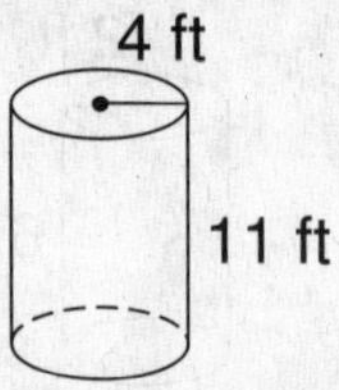

____

3.

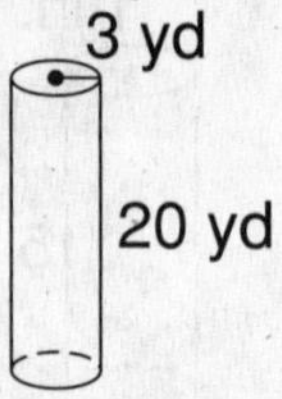

____

4.

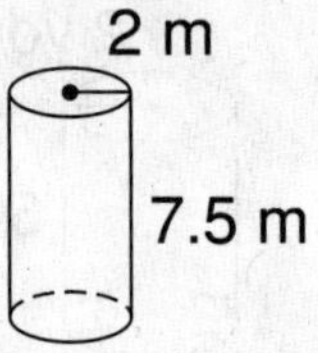

____

5.

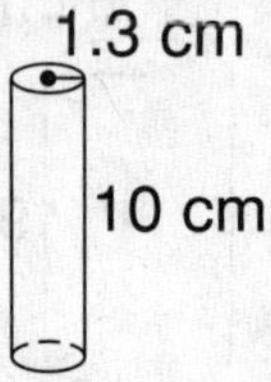

____

6.

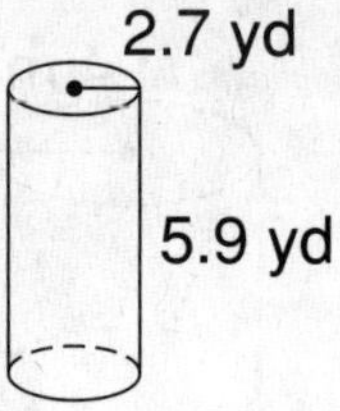

____

7.

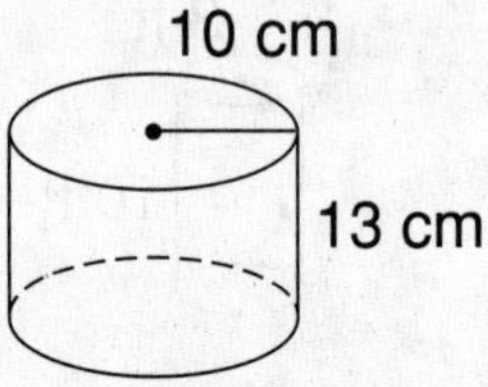

____

8.

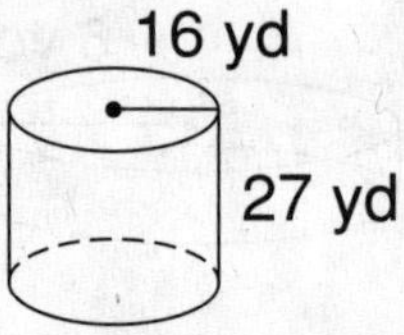

____

9.

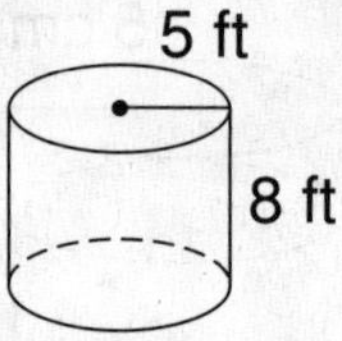

____

10. A cylindrical package of oatmeal is 20 centimeters tall. The diameter of its base is 10 centimeters. About how much oatmeal does the package hold?

____

11. The volume of a can is about 50.24 in^3. The radius of its base is 2 inches. How tall is the can?

____

Name ______________________ Date __________ Class __________

LESSON **10-8**

Practice C
Volume of Cylinders

Find the volume V of each cylinder to the nearest cubic unit.

1.

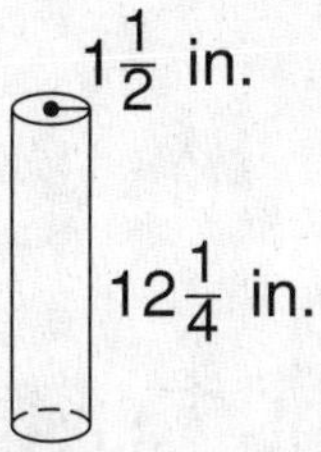

____

2.

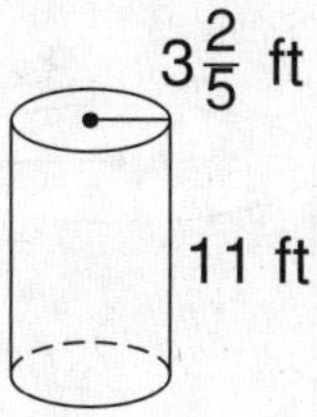

____

3.

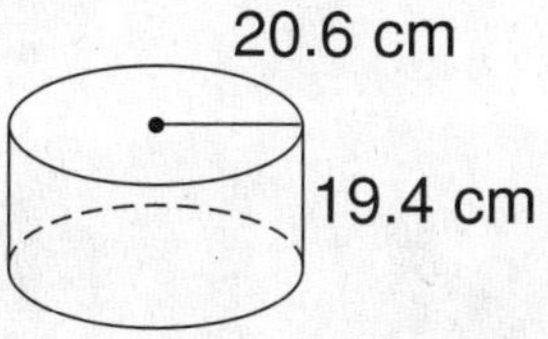

____

4.

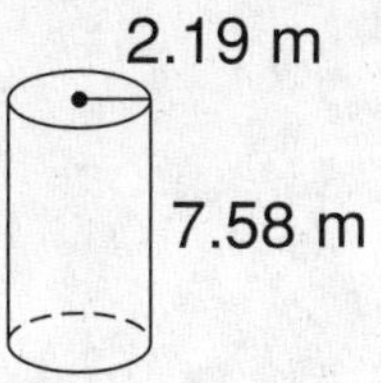

____

5.

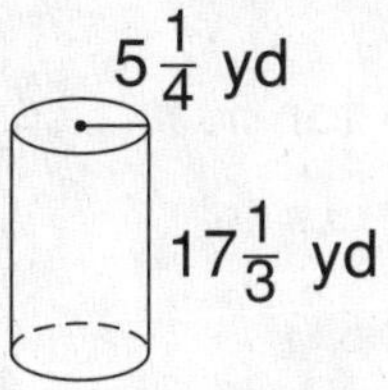

____

6.

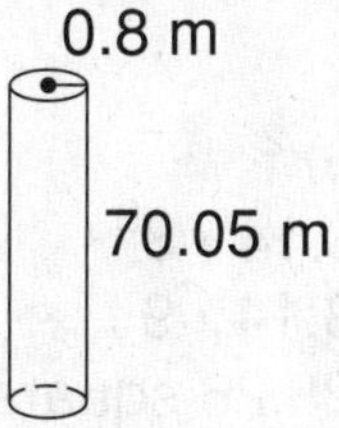

____

Find the volume of each cylinder using the information given.

7. $d = 39.2$ cm; $h = 48.5$ cm

____

8. $r = 8\frac{1}{2}$ in.; $h = 25$ in.

____

9. $r = 15$ m; $h = 68$ m

____

10. $d = 12$ ft; $h = 11.7$ ft

____

11. The base of a cylinder has a circumference of 25.12 inches. The height of the cylinder is 10 inches. What is the volume of the cylinder?

____

12. The height of a cylinder is 3 times greater than the diameter of its base. The area of the base is 78.5 cm^2. What is the volume of the cylinder? Use 3.14 for π.

____

Name ______________________ Date ____________ Class ____________

LESSON **10-8**

Reteach

Volume of Cylinders

You can use what you know about area to help you find the volume of a cylinder.

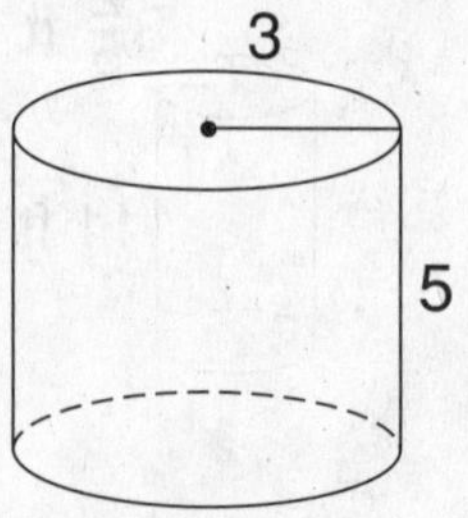

To find the volume of a cylinder, first find the area of the circular base.

$A = \pi r^2$ Use 3.14 for π.

$\approx 3.14 \cdot 3^2$

$\approx 3.14 \cdot 9$

≈ 28.26 square units

The area of the circular base is about 28.26 square units.

Next, multiply your answer by the height of the cylinder. The height of the cylinder is 5 units, so multiply 28.26 by 5.

$28.26 \cdot 5 = 141.3$

So, the volume of the cylinder is about 141.3 cubic units.

Find the volume of each cylinder.

1.

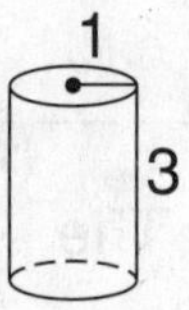

____

2.

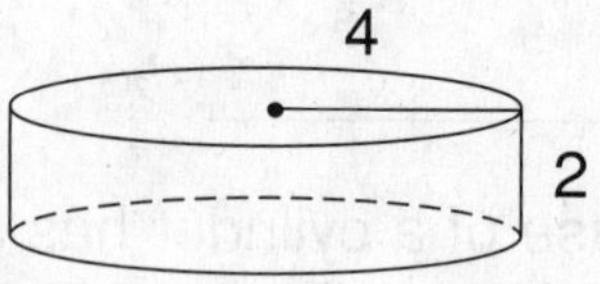

____

3.

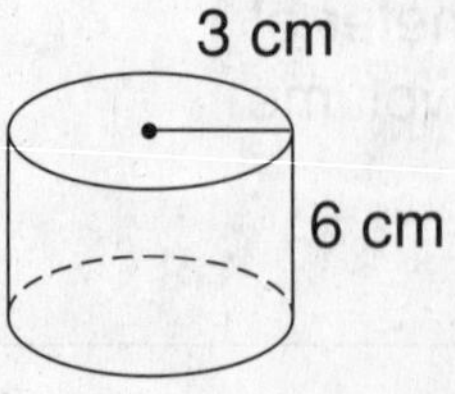

____

4.

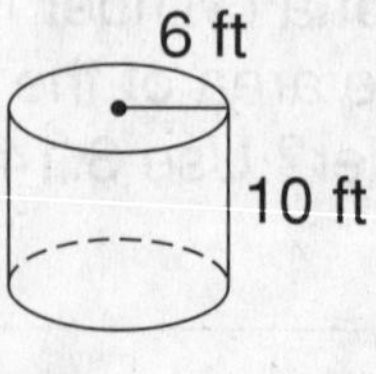

____

Name ______________________ Date ____________ Class ____________

LESSON 10-8

Challenge

Champion Trees

Trees are among Earth's largest and oldest living things. The National Register of Big Trees measures trees throughout the United States to find the largest of each species—the champions. The four huge trees described below are the only champions that have been listed in the register since the program began in 1940.

Use the register's data to find the circumference and volume of each champion American tree trunk to the nearest foot and cubic foot.

$h = 261$ ft
$r = 13.6$ ft

"General Sherman"
Giant Sequoia, California

Circumference: ____________

Volume: ____________

$h = 40$ ft
$r = 3.3$ ft

"Jardine Juniper"
Rocky Mountain Juniper, Utah

Circumference: ____________

Volume: ____________

$h = 86$ ft
$r = 6.4$ ft

"Bennett Juniper"
Western Juniper, California

Circumference: ____________

Volume: ____________

$h = 96$ ft
$r = 5.1$ ft

"Wye Oak"
White Oak, Maryland

Circumference: ____________

Volume: ____________

Name ______________________ Date ____________ Class ____________

LESSON **10-8**

Problem Solving

Volume of Cylinders

Write the correct answer.

1. The Hubble Space Telescope was launched into space in 1990. Shaped like a cylinder, the telescope is 15.9 meters long, with a diameter of 4.2 meters. To the nearest whole cubic meter, what is the volume of the Hubble Space Telescope?

____

2. The Living Color aquarium in Bermuda is the largest freestanding cylindrical aquarium in the Western Hemisphere. With a 10-foot diameter and an 18-foot height, the aquarium holds 10,400 gallons of water! What is the aquarium's volume in cubic feet?

____

3. In 1902 an American music company built the world's largest music recording cylinder. Nicknamed "Brutus," the cylinder is 5 feet tall, with a 2-foot diameter. What is the volume of the "Brutus" cylinder?

____

4. The world's largest glass of orange juice was filled in Florida in 1998. At 8 feet tall and with a 2-foot radius, the glass held about 700 gallons of orange juice. What was the volume of that huge glass of orange juice?

____

Circle the letter of the correct answer.

5. A large can of soda is 7.5 inches tall and has a 3-inch diameter. A small can of soda is 5 inches tall with a 2.5-inch diameter. To the nearest cubic inch, how much more soda does the large can hold?

A 53 in^3 more soda

B 28 in^3 more soda

C 25 in^3 more soda

D 20 in^3 more soda

6. A cylindrical candle is tightly packed in a rectangular box with a volume of 144 in^3. Which of these could be the dimensions of the candle?

F $h = 6$ in.; $r = 3$ in.

G $h = 2$ in.; $r = 5$ in.

H $h = 4$ in.; $r = 3$ in.

J $h = 3$ in.; $r = 4$ in.

7. The maximum length for an official professional baseball bat is 36 inches. Its maximum diameter is 2.6 inches. To the nearest cubic inch, what is the maximum volume of a professional baseball bat?

A 21 in^3

B 119 in^3

C 191 in^3

D 764 in^3

8. A can of tennis balls is 21 centimeters tall and has a diameter of 8 centimeters. What is the volume of the tennis ball can?

F 17,408.16 cm^3

G 1,055.04 cm^3

H 527.52 cm^3

J 263.76 cm^3

Name ______________________ Date ____________ Class ____________

LESSON 10-8

Reading Strategies

Analyze Information

The volume of a cylinder is measured in cubic units.

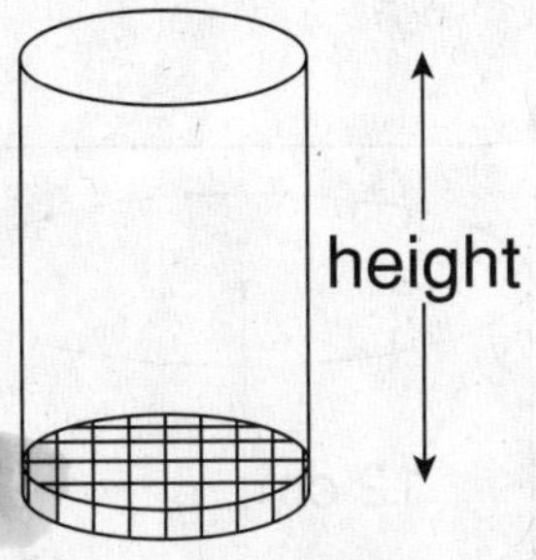

The area of the base gives the cubic units in the first layer. →	*Area of circular base*
The height shows the number of layers. →	*Area of base • height*
The base multiplied by the height is the volume of the cylinder. →	*V = area of base • height*

1. What measurement shows the number of cubic units in the first layer of a cylinder?

____

2. What measurement shows the number of layers in a cylinder?

____

3. How do you find the volume of a cylinder?

____

____

4. How do you find the area of the base of a cylinder?

____

Name ______________________ Date ____________ Class ____________

LESSON **10-8**

Puzzles, Twisters & Teasers

Ice Ages

Find the volume of each cylinder. Match the letters to the correct answers to solve the riddle. Use 3.14 for π and round to the nearest tenth.

1. S ______________

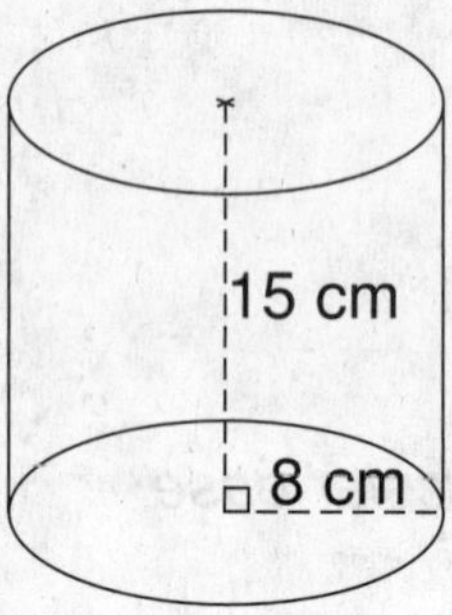

2. R ______________

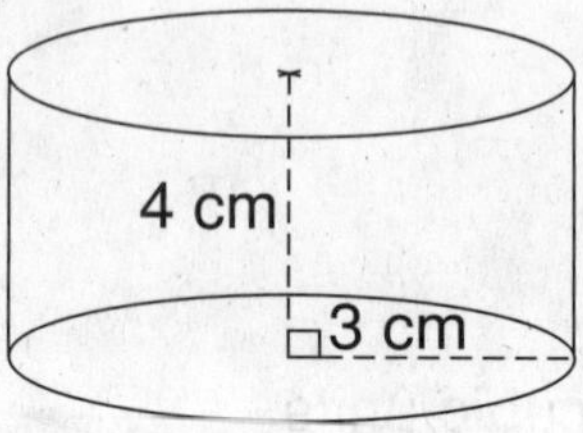

3. O ______________

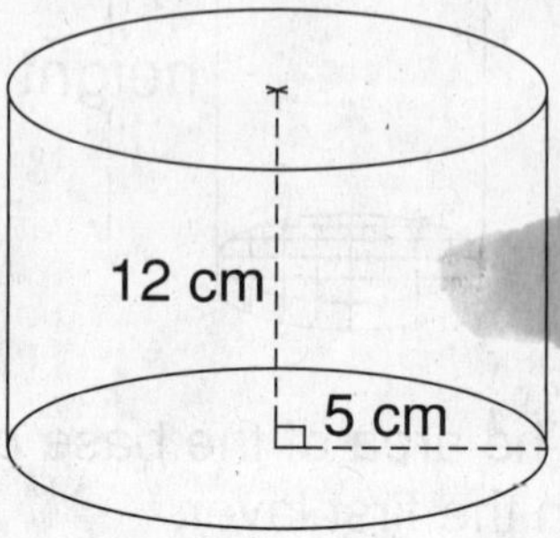

4. T ______________

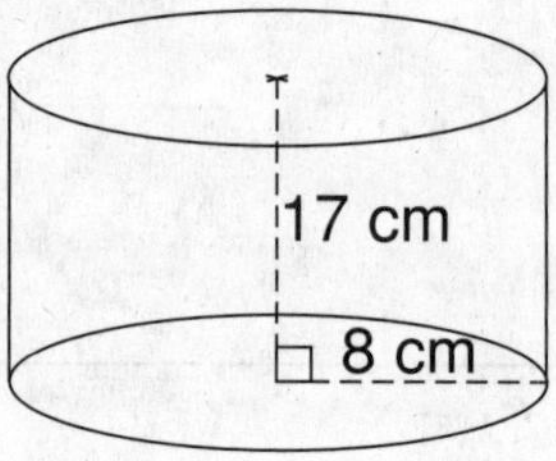

5. U ______________

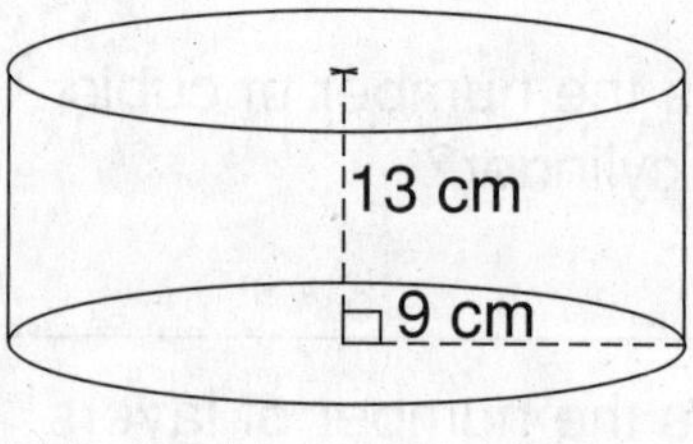

6. P ______________

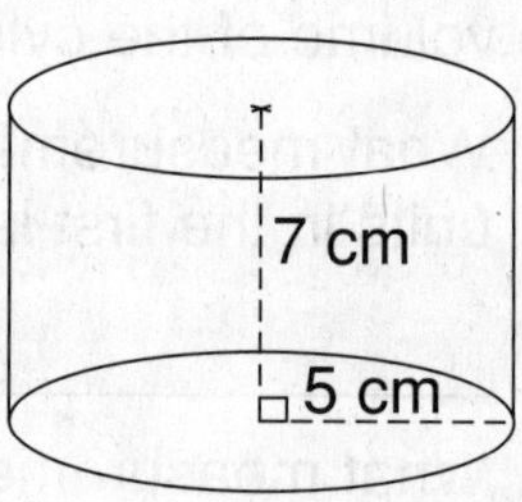

7. N ______________

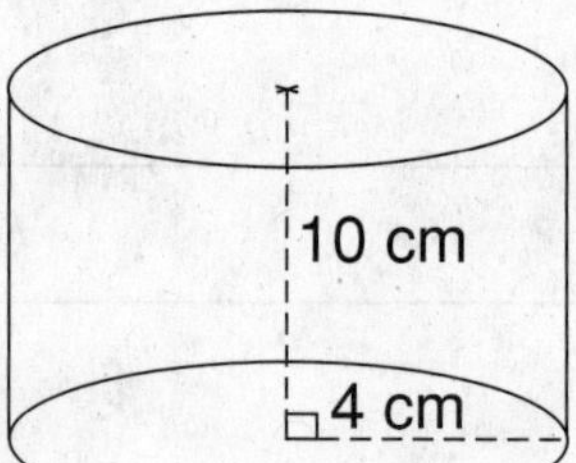

8. A ______________

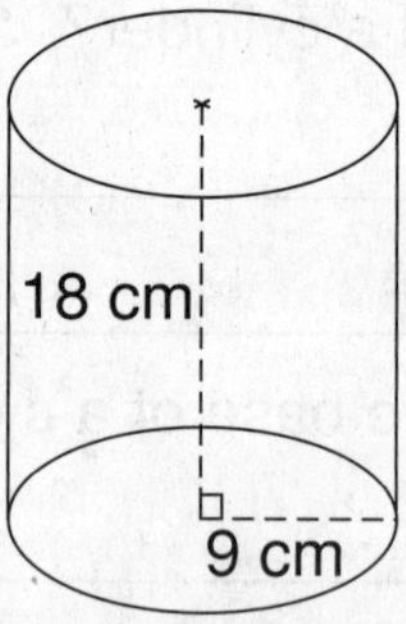

How does one dinosaur tell another to hurry up?

____	____	____	____	____	____
549.5	113	942	502.4	3,416.3	942

____	____	____	____	____	____
3,014.4	4,578.1	3,306.4	113	3,306.4	3,014.4

Name ______________________ Date __________ Class __________

LESSON 10-9

Practice A
Surface Area

Find the surface area *S* of each net.

1.

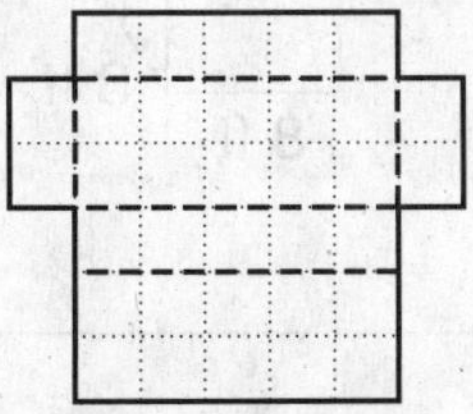

□ = 1 ft^2

____

2.

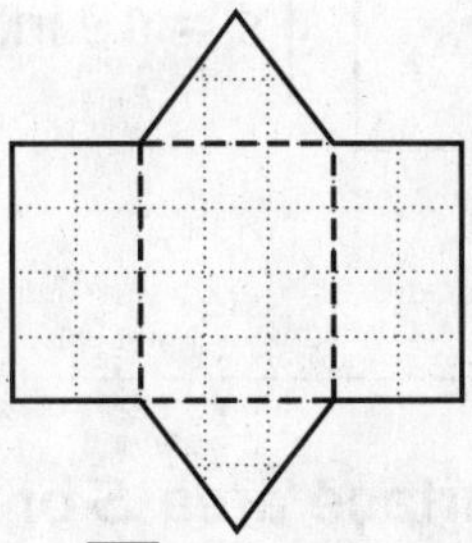

□ = 1 in^2

____

3.

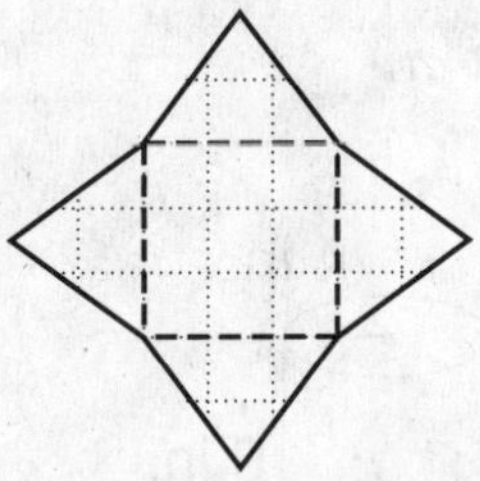

□ = 1 yd^2

____

4.

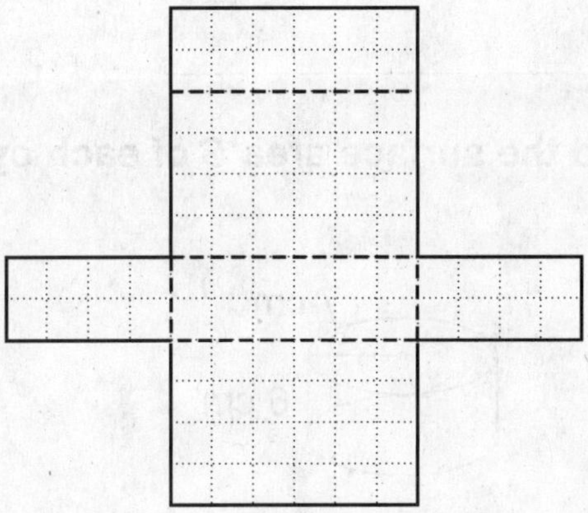

□ = 1 m^2

____

Find the surface area *S* of each prism.

5.

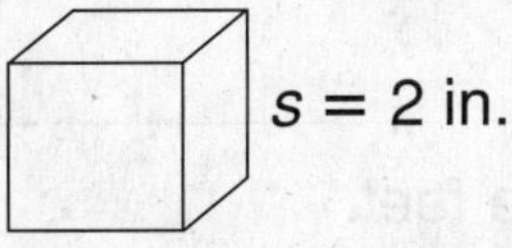

____

6.

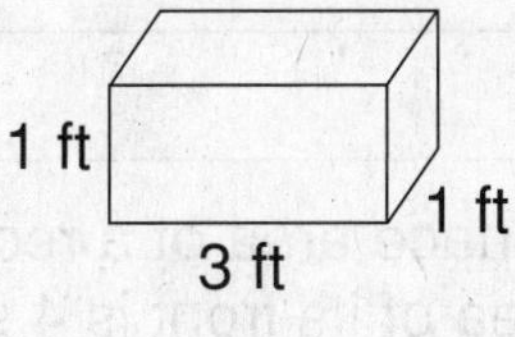

____

Name ______________________ Date __________ Class __________

LESSON 10-9

Practice B

Surface Area

Find the surface area *S* of each prism.

1.

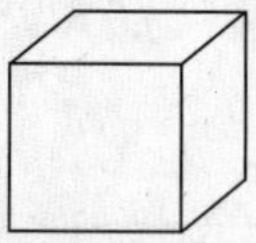

____

2.

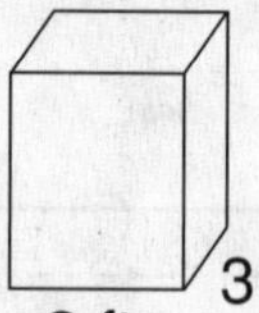

____

Find the surface area *S* of each pyramid.

3.

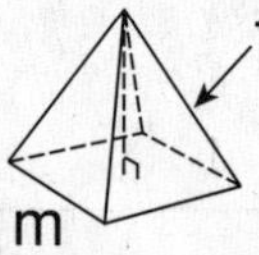

____

4.

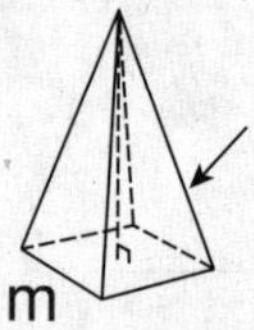

____

Find the surface area *S* of each cylinder. Use 3.14 for π.

5.

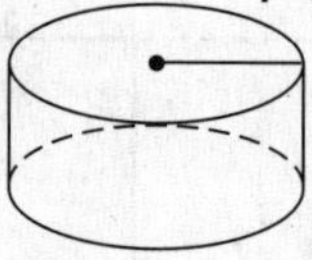

____

6.

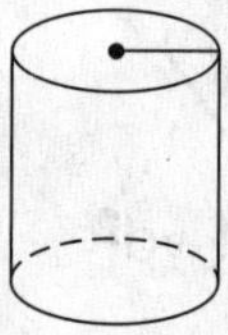

____

7. Why can you find an exact surface area measurement for a prism and pyramid but not for a cylinder?

____

____

8. The surface area of a rectangular prism is 48 square feet. The area of its front is 4 square feet, and the area of one side is 10 square feet. What is the area of the top of the prism?

____

Name ______________________ Date __________ Class __________

LESSON 10-9

Practice C

Surface Area

Find the surface area *S* of each prism.

1.

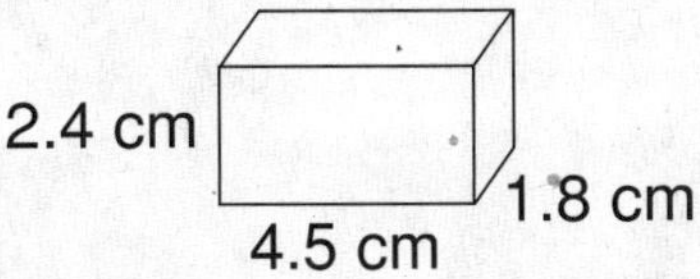

____

2.

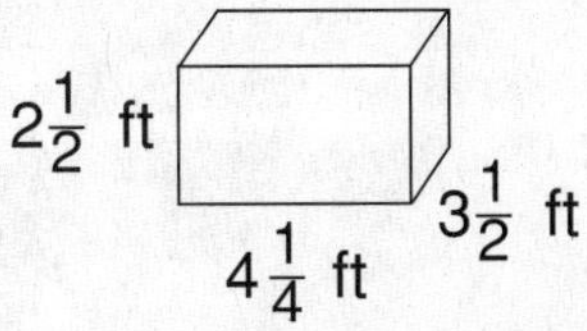

____

Find the surface area *S* of each pyramid.

3.

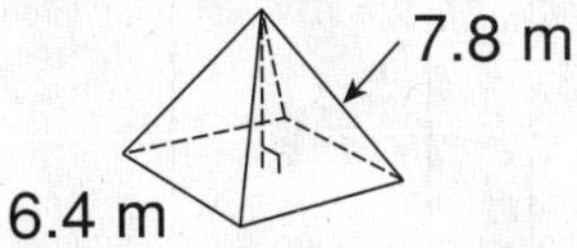

____

4.

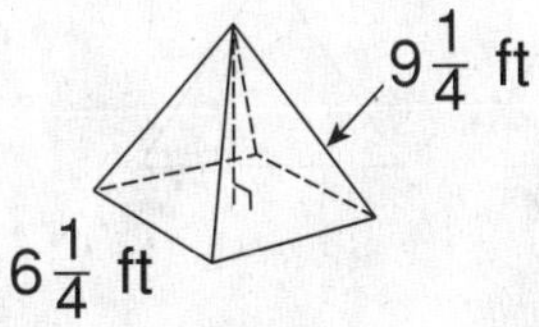

____

Find the surface area *S* of each cylinder. Use 3.14 for π, and round to the nearest hundredth.

5.

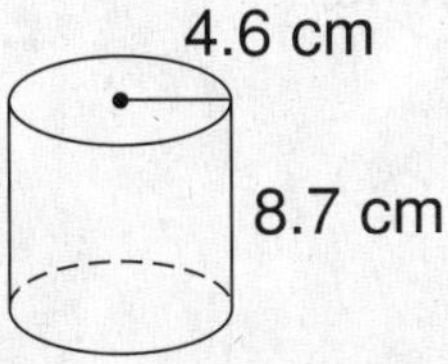

____

6.

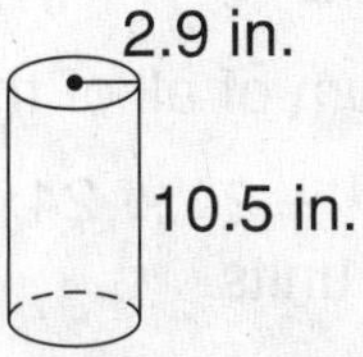

____

7. A cylindrical shipping box has a diameter of 4 inches and a height of 2 feet. What is the surface area of the box?

____

8. A rectangular box has no top. It is 6 inches long, 4 inches wide, and 5 inches tall. What is the surface area of the box?

____

Name ______________________ Date ____________ Class ____________

LESSON **10-9**

Reteach
Surface Area

You can use what you know about finding the area of polygon to find the surface area of a solid figure.

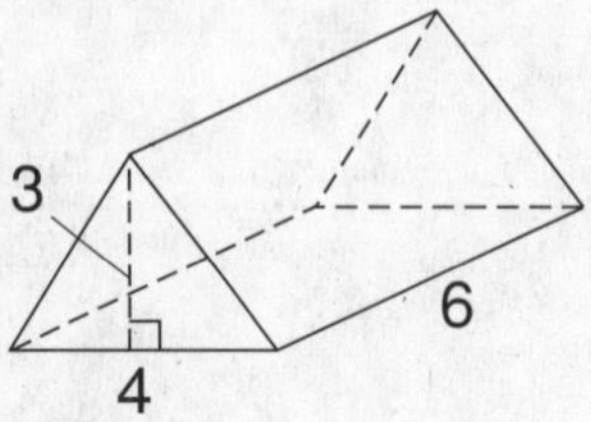

To find the surface area of the prism above, first find the area of each face.

2 congruent triangular bases

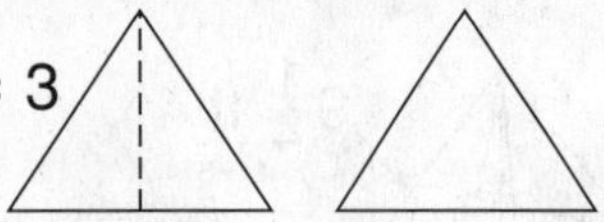

3 rectangular faces

$A = \frac{1}{2}bh$

$= \frac{1}{2}(4 \cdot 3)$

$= 6$ square units

$A = \ell w$

$= 4 \cdot 6$

$= 24$ square units

Then, find the sum of all of the faces of the prism.

$SA = 6 + 6 + 24 + 24 + 24$

$= 84$ square units

Find the surface area of each figure.

1.

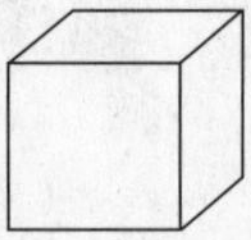

____

2.

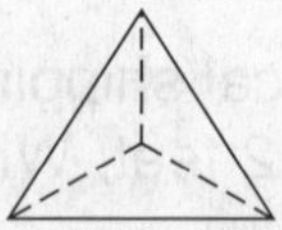

____

Name ______________________________ Date ____________ Class ____________

LESSON 10-9

Challenge

A Monumental Paint Job

Completed in 1927, the Lincoln Memorial in Washington, D.C., honors President Abraham Lincoln. The monument was built in the style of Greek temples, with 36 columns surrounding the outside of the monument and 8 columns surrounding the statue of Lincoln inside.

It's your job to paint the columns of the Lincoln Memorial. You will paint its exterior columns blue, and its interior columns red. Use the diagram to figure out how much paint you should buy and how much the paint will cost.

each outside column:
$h = 44$ ft
$r = 3.7$ ft

each inside column:
$h = 50$ ft
$r = 3.75$ ft

Surface Areas to Paint	Gallons of Paint Needed	Cost of Paint
Remember to consider the parts of each column that will **not** be painted.	One gallon of paint will cover 350 square feet. Round **up** to the nearest whole gallon.	Each gallon of paint costs $16.99.
Exterior Columns:	Exterior Columns:	Blue Paint:
Interior Columns:	Interior Columns:	Red Paint:

Name ______________________________ Date ____________ Class ____________

LESSON 10-9

Problem Solving

Surface Area

Write the correct answer.

1. The world's largest cookie was baked in Wisconsin in 1992. Its diameter was 34 feet and contained about 4 million chocolate chips! If the cookie was a cylinder 1 foot tall, and you wanted to cover it with icing, how many square inches would you have to ice? Use 3.14 for π.

____

2. The top of the Washington Monument is a square pyramid covered with white marble. Each triangular face is 58 feet tall and 34 feet wide. About how many square feet of marble covers the top of the monument? (The base is hollow.)

____

3. The Parthenon, a famous temple in Greece, is surrounded by large stone columns. Each column is 10.4 meters tall and has a diameter of 1.9 meters. To the nearest whole square meter, what is the surface area of each column (not including the top and bottom)?

____

4. The tablet that the Statue of Liberty holds is 7.2 meters long, 4.1 meters wide, and 0.6 meters thick. The tablet is covered with thin copper sheeting. If the tablet was freestanding, how many square meters of copper covers the statue's tablet?

____

Circle the letter of the correct answer.

5. The largest Egyptian pyramid is called the Great Pyramid of Khufu. It has a 756-foot square base and a slant height of 481 feet. What is the total surface area of the faces of the Pyramid of Khufu?

A 727,272 ft^2 **C** 727,727 ft^2

B 727,722 ft^2 **D** 772,272 ft^2

6. A glass triangular prism for a telescope is 5.5 inches tall. Each side of the triangular base is 4 inches long, with a 3-inch height. How much glass covers the surface of the prism?

F 6 in^2 **H** 39 in^2

G 12 in^2 **J** 78 in^2

7. A can of frozen orange juice is 7.5 inches tall, and its base diameter is 3.5 inches. What size strip of paper is used for its label?

A 82.43 in^2 **C** 576.98 in^2

B 26.25 in^2 **D** 101.66 in^2

8. Tara made fuzzy cubes to hang in her car. Each side of the 2 cubes is 4 inches long. How much fuzzy material did Tara use to make both cubes?

F 96 in^2 **H** 16 in^2

G 192 in^2 **J** 128 in^2

Name ______________________ Date __________ Class __________

LESSON 10-9

Reading Strategies

Follow a Procedure

Surface area is the sum of the areas of all the faces of a solid figure.

If you unfold a solid figure and lay it flat, you make a net. A **net** is a pattern that makes it easy to see all faces of a solid figure.

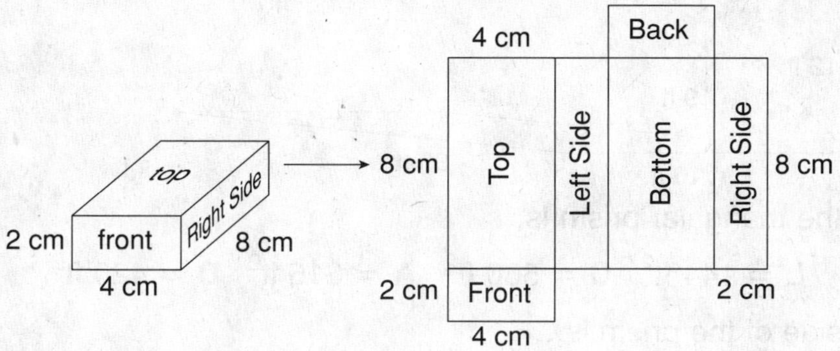

Follow these steps to find the surface area of this rectangular prism. Use the formula $A = \ell w$ to find the area of each face.

Step 1: Find the area of the top face. The bottom face has the same area.

↓

Step 2: Find the area of the right face. The left face has the same area.

↓

Step 3: Find the area of the front face. The back face has the same area.

↓

Step 4: Add the areas of all six faces to find the total surface area.

Answer the following questions.

1. What is the area of the top face? What face has the same area?

____

2. What is the area of the right face? What face has the same area?

____

3. What is the area of the front face? What face has the same area?

____

4. What numbers do you add to find the total surface area of the prism? What is the total surface area?

____

Name ______________________ Date __________ Class __________

LESSON **10-9**

Puzzles, Twisters & Teasers

See Saw

For problems 1–4 use the triangular prism to answer each question. For problems 5–8 use the rectangular prism to answer each question. Choose the correct answer and write the letter in the space matching the problem number to solve the riddle.

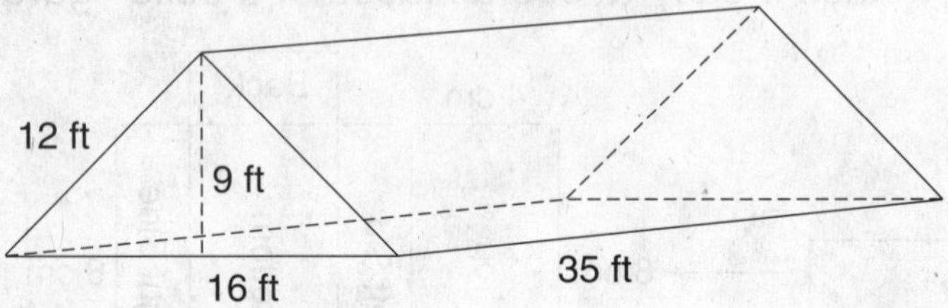

1. The area of the bottom of the triangular prism is…

W = 37 ft^2 **R** = 72 ft^2 **L** = 144 ft^2 **C** = 560 ft^2 **A** = 315 ft^2 **D** = 420 ft^2

2. The area of the triangular side of the prism is…

T = 37 ft^2 **O** = 72 ft^2 **E** = 144 ft^2 **L** = 560 ft^2 **B** = 315 ft^2 **S** = 420 ft^2

3. The area of the back face of the triangular prism is…

D = 37 ft^2 **N** = 72 ft^2 **J** = 144 ft^2 **P** = 560 ft^2 **E** = 315 ft^2 **A** = 420 ft^2

4. The total surface area of the triangular prism is…

M = 840 ft^2 **T** = 984 ft^2 **D** = 1,472 ft^2 **I** = 1,544 ft^2

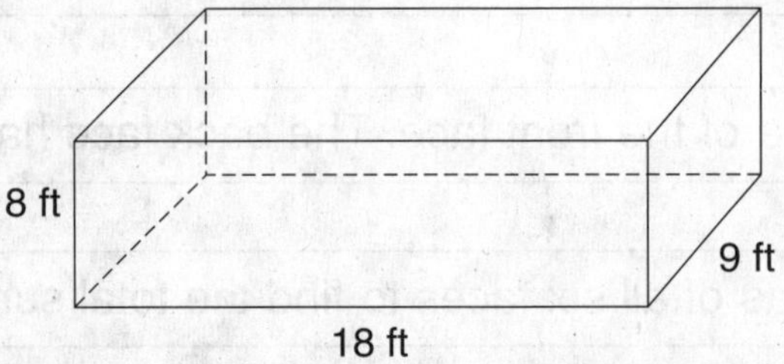

5. The area of the front face is…

M = 52 ft^2 **A** = 72 ft^2 **S** = 162 ft^2 **N** = 144 ft^2 **T** = 172 ft^2 **L** = 1,296 ft^2

6. The area of the right-hand face is…

C = 52 ft^2 **E** = 72 ft^2 **K** = 162 ft^2 **A** = 144 ft^2 **B** = 172 ft^2 **D** = 1,296 ft^2

7. The area of the bottom face is …

T = 52 ft^2 **A** = 72 ft^2 **L** = 162 ft^2 **O** = 144 ft^2 **G** = 172 ft^2 **C** = 1,296 ft^2

8. The total surface area of this rectangular prism is…

R = 756 ft^2 **B** = 840 ft^2 **N** = 942 ft^2 **I** = 1,544 ft^2

What puts white lines on the ocean?

___ ___   ___ ___ ___ ___ ___   ___ ___ ___ ___ ___

3 5   2 1 6 3 5   7 4 5 6 8

LESSON 10-1 Practice A
Estimating and Finding Area

Estimate the area of each figure.

1. 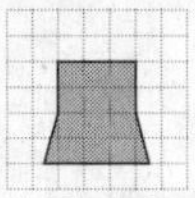

■ $= 1\ ft^2$

about 13 square feet

2. 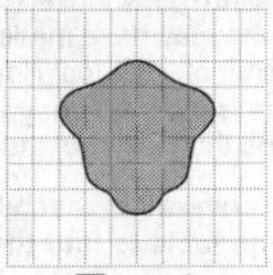

■ $= 1\ in^2$

about 24 square inches

3. 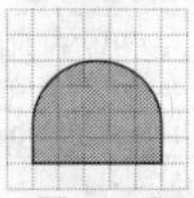

■ $= 1\ m^2$

about 18 square meters

4. 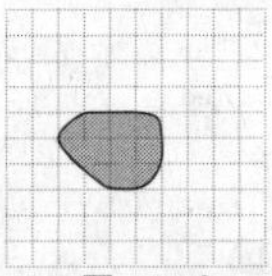

■ $= 1\ yd^2$

about 9 square yards

Find the area of each rectangle.

5. 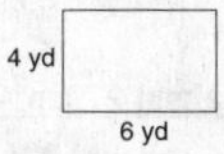

$24\ yd^2$

6. 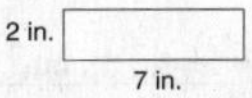

$14\ in^2$

7. 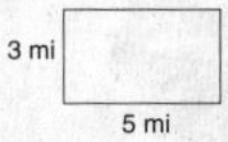

$15\ mi^2$

8. 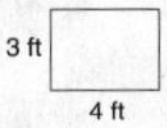

$12\ ft^2$

9. A square room has sides that each measure 5 feet. How many square feet of carpet is needed to cover the room's entire floor?

$25\ ft^2$

10. A rectangular coffee table is 2 feet wide and 4 feet long. How many square feet of glass is needed to cover the entire table top?

$8\ ft^2$

LESSON 10-1 Practice B
Estimating and Finding Area

Estimate the area of each figure.

1. 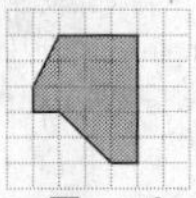

■ $= 1\ ft^2$

about 15 square feet

2. 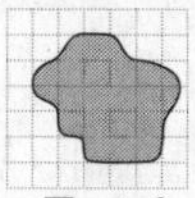

■ $= 1\ m^2$

about 19 square meters

Find the area of each rectangle.

3. 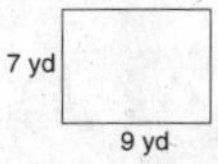

$63\ yd^2$

4. 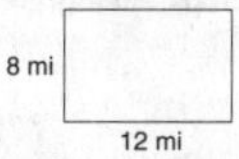

$96\ mi^2$

Find the area of each parallelogram.

5. 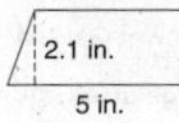

$10.5\ in^2$

6. 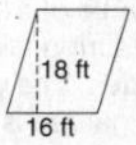

$288\ ft^2$

7. Mariah is planting a rectangular rose garden. In the center of the garden, she puts a smaller rectangular patch of grass. The grass is 2 ft by 3 ft. What is the area of the rose garden?

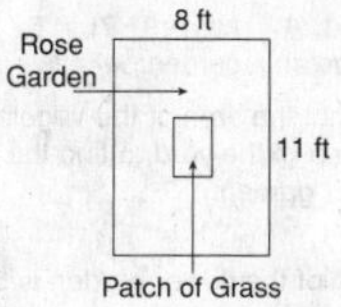

$82\ ft^2$

8. A section of a stained-glass window is shaped like a parallelogram. Its base is 6.5 inches, and its height is 4 inches. How much glass is needed to cover the section completely?

$26\ in^2$

9. Your rectangular yard is 10 feet wide and 26 feet long. How many square feet of grass do you need to plant if you want to cover the entire yard?

$260\ ft^2$

LESSON 10-1 Practice C
Estimating and Finding Area

Estimate the area of each figure.

1. 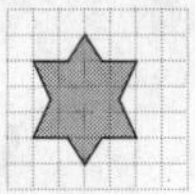

■ $= 1\ ft^2$

about 11 square feet

2. 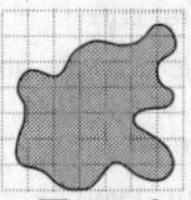

■ $= 1\ m^2$

about 25 square meters

Find the area of each rectangle.

3. 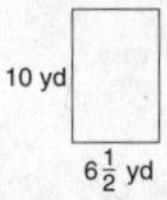

$65\ yd^2$

4. 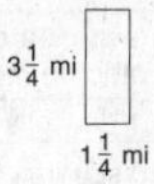

$4\frac{1}{16}\ mi^2$

Find the area of each parallelogram.

5. 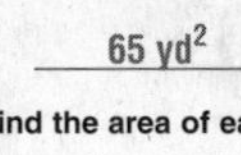

$337.68\ in^2$

6. 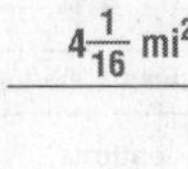

$5\frac{5}{6}\ ft^2$

7. Denzel is building a square deck. On one side of the deck, he puts a smaller square hot tub. The hot tub is 5 ft long. What is the area of the deck?

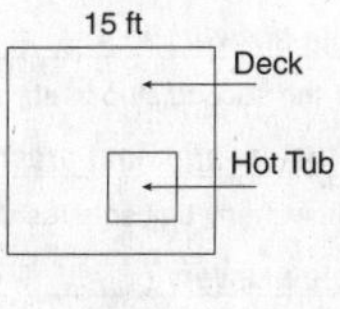

$200\ ft^2$

8. A kite is in the shape of a parallelogram with a 4.5 cm base and a height of 7.9 cm. What is the area of the kite?

$35.55\ cm^2$

LESSON 10-1 Reteach
Estimating and Finding Area

You can use grid paper to estimate and find area of figures. The area of a figure is the amount of surface it covers. Area is measured in square units.

To estimate the area of the figure below, first find the number of fully shaded squares. Next, find the number of nearly full squares. Then, find the number of half or nearly half-shaded squares.

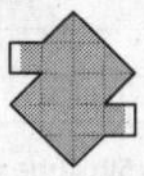

There are 6 fully shaded squares.
There are 2 nearly full squares.
There are 8 half-shaded squares.
$(8 \cdot \frac{1}{2} = 4)$

Find the estimated total number of squares shaded.

$6 + 2 + 4 = 12$. The area of the figure is about 12 square units.

Estimate the area of each figure. Possible answers:

1. 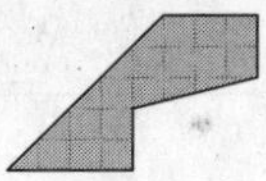

18 units

2. 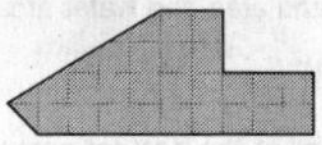

27 units

To find the area of a rectangle, find the total number of square units.

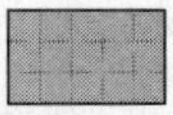

There are 3 rows of 5 squares.
$3 \cdot 5 = 15$
So, the area of the rectangle is 15 square units.

Find the area of each rectangle.

3. 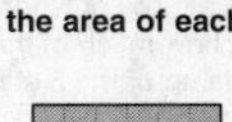

12 square units

4. 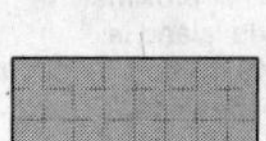

32 square units

LESSON 10-1
Reteach
Estimating and Finding Area (continued)

To find the area of a parallelogram, first turn your parallelogram into a rectangle.

Then find the area of the rectangle.

Because the area of a rectangle is $A = \ell w$, the area of a parallelogram is $A = bh$.

$4 \cdot 2 = 8$

So, the area of the parallelogram is 8 square units.

Find the area of each parallelogram.

5. 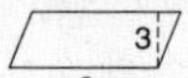

24 square units

6. 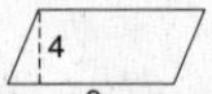

36 square units

Sometimes you need to subtract a smaller area from a larger area.

A rectangular yard is made up of a rectangular flower garden surrounding a rectangular vegetable garden. The yard is 9 ft by 7 ft. The vegetable garden is 3 ft by 2 ft. What is the area of the flower garden?

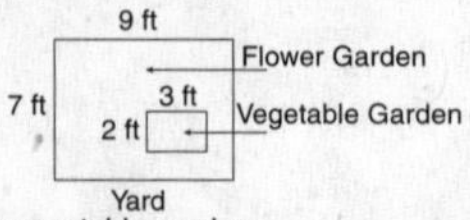

First find the area of the yard and the area of the vegetable garden.

Area of yard: $A = lw = (9 \cdot 7) = 63$
Area of vegetable garden: $A = lw = (3 \cdot 2) = 6$

Then subtract the area of the vegetable garden from the area of the yard to find the area of the flower garden.

$63 - 6 = 57$
So the area of the flower garden is 57 square feet.

Solve.

7. A rectangular yard is made up of a rectangular grass lawn surrounding a rectangular patio. The yard is 12 ft by 10 ft. The patio is 5 ft by 4 ft. What is the area of the grass lawn?

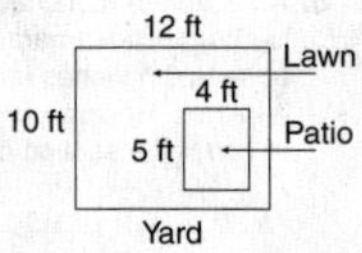

100 square feet

LESSON 10-1
Challenge
Chewing Gum Archaeology

Archaeologists found the oldest piece of chewing gum in Sweden—it was 9,000 years old! The scientists studied the gum to learn about the people who may have made and chewed it. You can be a chewing gum archaeologist, too. How? When a standard-sized wad of chewed gum is stepped on, you can study the area of the flattened gum to find the weight of the person who squished it!

Estimate the area covered by each flattened piece of gum below. Then use the table to find the gum squisher's weight. Possible answers:

Bubble gum was invented in Pennsylvania in 1928.

Estimated Area of Squished Gum	Estimated Weight of Gum Squisher
about 25 cm^2	about 50 lb
about 50 cm^2	about 100 lb
about 75 cm^2	about 150 lb
about 100 cm^2	about 200 lb

1. 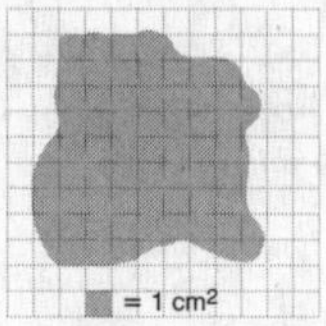

Estimated Area: about 65 cm^2
Estimated Weight: about 150 lb

2. 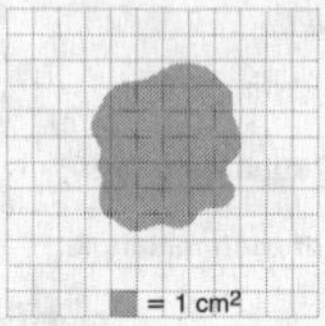

Estimated Area: about 27 cm^2
Estimated Weight: about 50 lb

3. 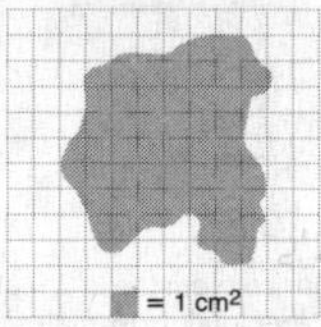

Estimated Area: about 50 cm^2
Estimated Weight: about 100 lb

4. 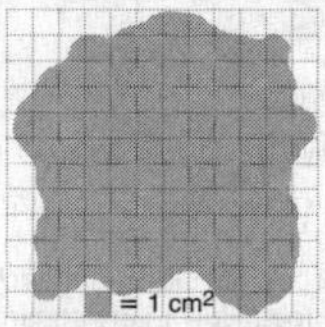

Estimated Area: about 105 cm^2
Estimated Weight: about 200 lb

LESSON 10-1
Problem Solving
Estimating and Finding Area

Use the table to answer each question.

State Information

State	Approx. Width (mi)	Approx. length (mi)	Water Area (mi^2)
Colorado	280	380	376
Kansas	210	400	462
New Mexico	343	370	234
North Dakota	211	340	1,724
Pennsylvania	160	283	1,239

1. New Mexico is the 5th largest state in the United States. What is its approximate total area?

126,910 mi^2

2. Kansas is the 15th largest state in the United States. What is its approximate total area?

84,000 mi^2

3. What is the difference between North Dakota's land area and water area?

70,016 mi^2

4. What is Pennsylvania's approximate land area?

45,280 mi^2

Circle the letter of the correct answer.

5. What is the difference between Colorado's land area and Pennsylvania's land area?
   - **A** 106,400 mi^2
   - **(B)** 61,120 mi^2
   - **C** 60,120 mi^2
   - **D** 45,280 mi^2

6. About what percent of the total area of Pennsylvania is covered by land?
   - **F** about 3%
   - **G** about 30%
   - **H** about 67%
   - **(J)** about 97%

7. Rhode Island is the smallest state. Its total land area is approximately 1,200 mi^2. Rhode Island is approximately 40 miles long. About how wide is Rhode Island?
   - **A** about 20 mi
   - **B** about 40 mi
   - **C** about 50 mi
   - **(D)** about 30 mi

8. The entire United States covers 3,794,085 square miles of North America. About how much of that area is not made up of the 5 states in the chart?
   - **F** 2,537,470 mi^2
   - **(G)** 3,359,755 mi^2
   - **H** 3,686,525 mi^2
   - **J** 3,1310,818 mi^2

LESSON 10-1
Reading Strategies
Use a Flowchart

Area is the measure of the amount of surface a figure covers.

Area is measured in square units.

This pond has a grid placed over it so that the number of square units the pond covers can be estimated.

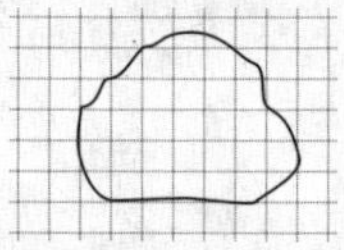

Follow this flowchart to help you estimate the area of the pond.

Step 1: Count full squares and almost full squares.
↓
Step 2: Count squares that are about half full. Add two half squares together to make a full square.
↓
Step 3: Do not count squares that are almost empty.
↓
Step 4: Add to estimate the total.

Answer the following questions.

1. What do you call the amount of surface a figure covers?

area

2. What is the first step in estimating the area of the pond above?

Count the number of full and almost full squares.

3. What is your estimate for the number of full squares?

Possible answer: 26

4. What is the second step in estimating the area of the pond?

Count the squares that are about half full.

5. About how many half squares did you count?

Possible answer: 7

6. What is your estimate for the area of the pond?

Possible answer: $29\frac{1}{2}$ square units

LESSON **10-1**

Puzzles, Twisters & Teasers

What's Furry?

Find the area of each figure below. Then find the letter matching the answer in the Answer Box. To solve the riddle, match the letter of the correct answer to the number of the problem.

1.

Area = 25 units^2

2.

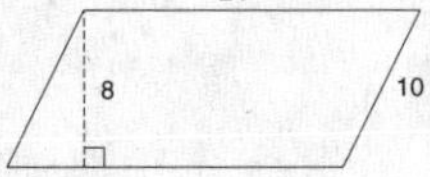

Area = 192 units^2

3.

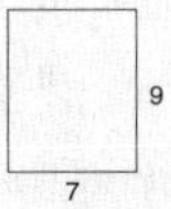

Area = 63 units^2

4.

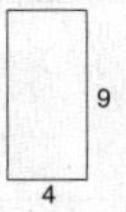

Area = 36 units^2

5.

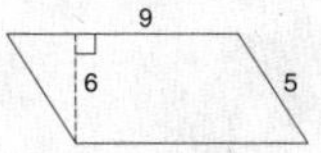

Area = 54 units^2

6.

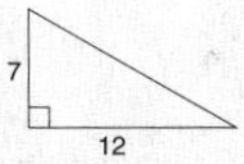

Area = 42 units^2

7.

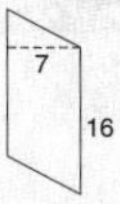

Area = 112 units^2

8.

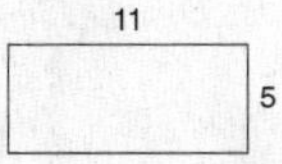

Area = 55 units^2

9.

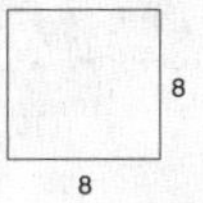

Area = 64 units^2

Answer Box

A = 112	D = 26	H = 42	K = 64	N = 55	R = 15
B = 28	E = 46	I = 63	L = 110	O = 54	T = 26
C = 25	G = 30	J = 240	M = 192	P = 36	W = 72

What do you call a monkey that sells potato chips?

A (7) C (1) H (6) I (3) P (4) M (2) O (5) N (8) K (9)

LESSON **10-2**

Practice A

Area of Triangles and Trapezoids

Complete. Write *triangle* or *trapezoid*.

1. The area of a triangle is half the product of its base and its height.
2. The area of a trapezoid is the product of half its height and the sum of its bases.

Find the area of each triangle.

3.

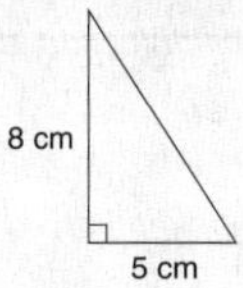

20 cm^2

4.

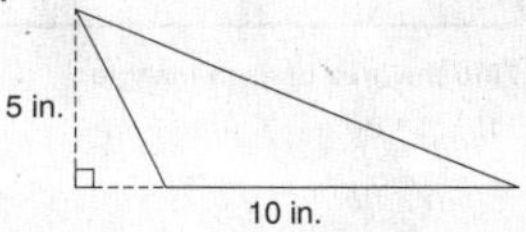

25 in^2

Find the area of each trapezoid.

5.

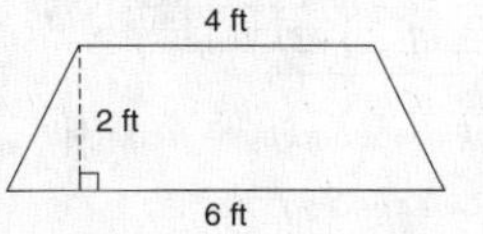

10 ft^2

6.

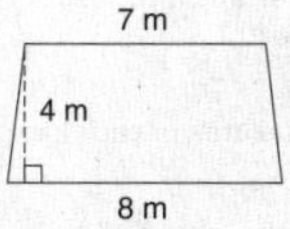

30 m^2

7. Andrew is making a triangular jewelry box. What is the area of the top of the box?

18 in^2

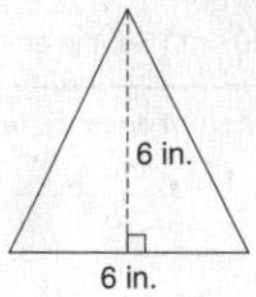

LESSON **10-2**

Practice B

Area of Triangles and Trapezoids

Find the area of each triangle.

1.

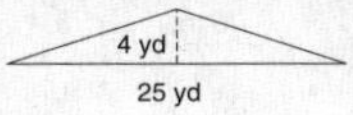

50 yd^2

2.

7 ft^2

3.

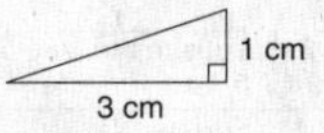

1.5 cm^2

4.

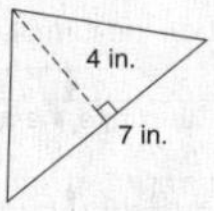

14 in^2

Find the area of each trapezoid.

5.

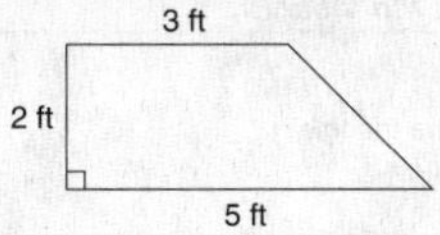

8 ft^2

6.

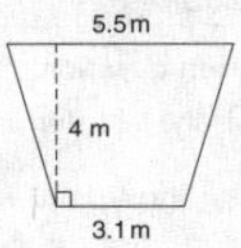

17.2 m^2

7.

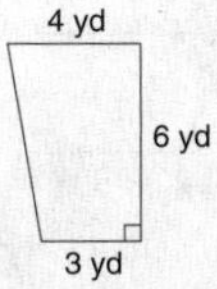

21 yd^2

8.

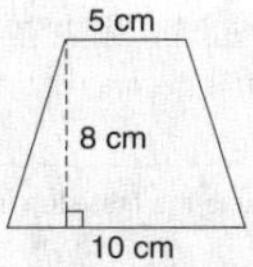

60 cm^2

9. The front part of a tent is 8 feet long and 5 feet tall. What is the area of the front part of the tent?

20 ft^2

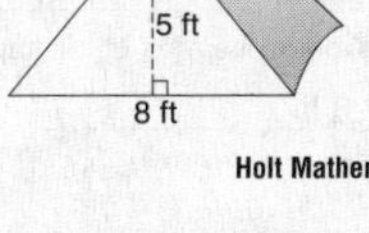

LESSON **10-2**

Practice C

Area of Triangles and Trapezoids

Find the area of each triangle.

1.

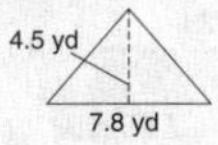

17.55 yd^2

2.

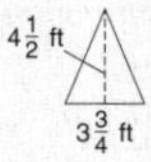

$8\frac{7}{16}$ ft^2

3.

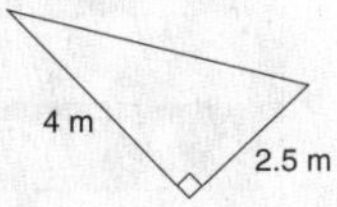

5 m^2

4.

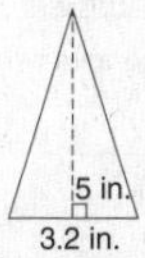

8 in^2

Find the area of each trapezoid.

5.

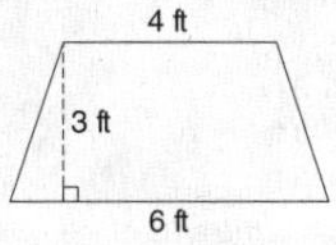

15 ft^2

6.

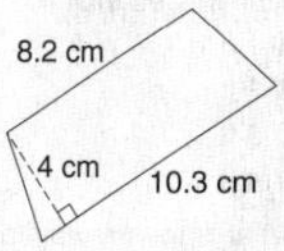

37 cm^2

7.

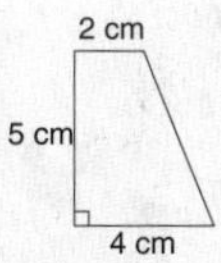

15 cm^2

8.

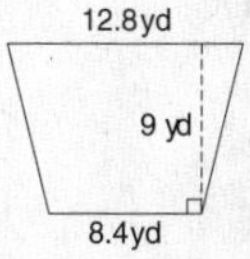

95.4 yd^2

9. A kite is formed by two congruent triangles with 6.2-cm bases and heights of 8.5 cm. What is the area of the entire kite?

52.7 cm^2

LESSON **10-2**

Reteach
Area of Triangles and Trapezoids

To find the area of a triangle, first turn your triangle into a rectangle.

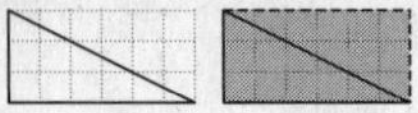

Next, find the area of the rectangle. 6 • 3 = 18

The triangle is half the area of the formed rectangle or $A = \frac{1}{2}bh$, so divide the product by 2.

18 ÷ 2 = 9 So, the area of the triangle is 9 square units.

Find the area of each triangle.

1. 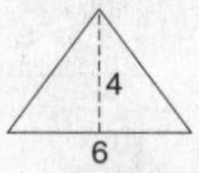

12 square units

2. 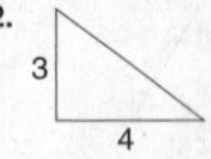

6 square units

To find the area of a trapezoid, first turn the trapezoid into two triangles and a rectangle.

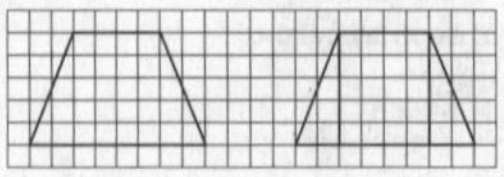

Find the area of each triangle.

$A = \frac{1}{2}bh$

$A = \frac{1}{2}(2 \cdot 5) = \frac{1}{2} \cdot 10 = 5$

Find the area of the rectangle.

$A = lw$

$A = 4 \cdot 5 = 20$

Now find the sum of the areas.

5 + 5 + 20 = 30 So, the area of the trapezoid is 30 square units.

Find the area of each trapezoid.

3. 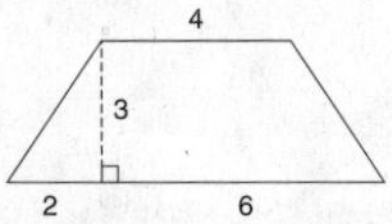

18 square units

4. 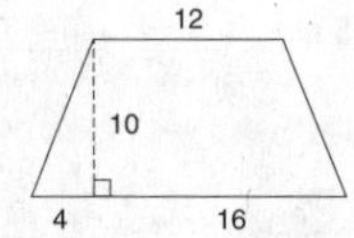

160 square units

LESSON **10-2**

Challenge
Pennant Problems

Find the amount of felt used to make each pennant.

1. 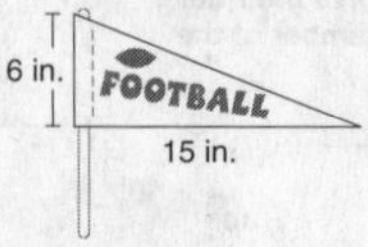

45 square inches

2. 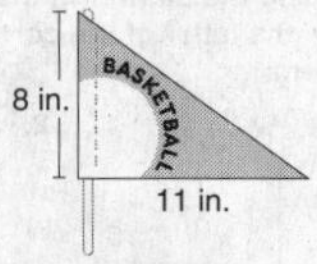

44 square inches

3. 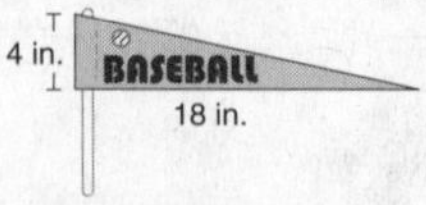

36 square inches

4. 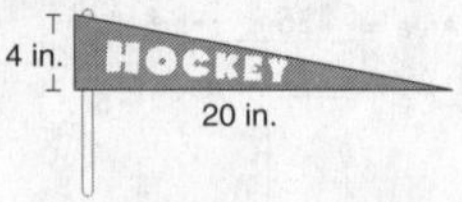

40 square inches

5. 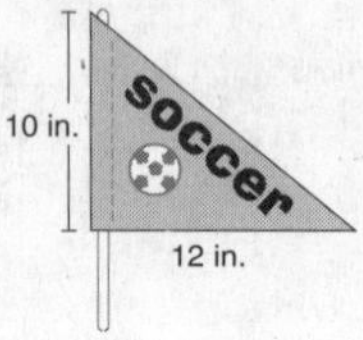

60 square inches

6. 

$42\frac{1}{2}$ square inches

7. List the pennants in order from least area to greatest area.

BASEBALL, HOCKEY, GOLF, BASKETBALL, FOOTBALL, SOCCER

LESSON **10-2**

Problem Solving
Area of Triangles and Trapezoids

Use the quilt design to answer the questions.

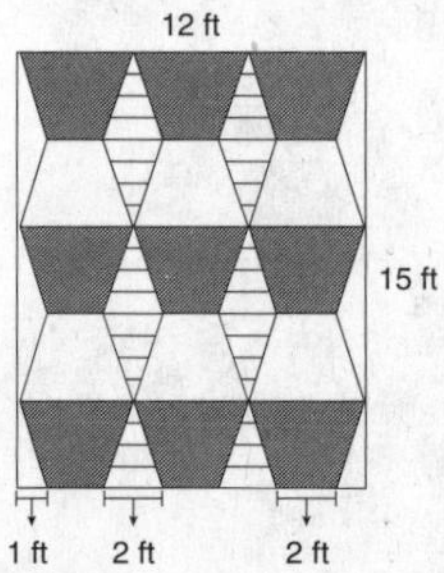

1. What are the lengths of the bases of each trapezoid?

4 ft and 2 ft

2. What is the height of each trapezoid?

3 ft

3. What is the area of each trapezoid?

9 ft²

Circle the letter of the correct answer.

4. What is the length of the base of each striped triangle?
   - A 1 ft
   - (B) 2 ft
   - C 3 ft
   - D 4 ft

5. What is the height of each striped triangle?
   - F 1 ft
   - G 2 ft
   - (H) 3 ft
   - J 5 ft

6. What is the area of each striped triangle?
   - (A) 3 ft²
   - B 1 ft²
   - C $\frac{3}{4}$ ft²
   - D $\frac{1}{4}$ ft²

7. What is the area of the quilt?
   - F 36 ft²
   - G 90 ft²
   - H 96 ft²
   - (J) 180 ft²

LESSON **10-2**

Reading Strategies
Follow a Procedure

Triangles and trapezoids are two different types of polygons. You can follow a procedure to help you find the area of each type of polygon.

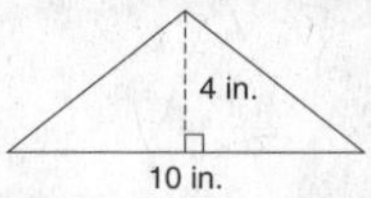

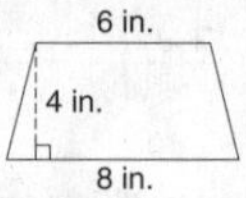

Triangle	Trapezoid
Step 1: Use the formula $A = \frac{1}{2}bh$.	Step 1: Use the formula $A = \frac{1}{2}h(b_1 + b_2)$.
Step 2: Substitute the length of the base for *b*.	Step 2: Substitute the length of the height for *h*.
Step 3: Substitute the length of the height for *h*.	Step 3: Substitute the lengths of the bases for b_1 and b_2 and add.
Step 4: Multiply.	Step 4: Multiply.

Answer each question.

1. What is the first step in finding the area of the triangle?

Use the formula $A = \frac{1}{2}bh$.

2. What are the second and third steps in finding the area of the triangle?

Substitute 10 for *b*. Substitute 4 for *h*.

3. What is the area of the triangle?

20 in²

4. What is the first step in finding the area of the trapezoid?

Use the formula $A = \frac{1}{2}h(b_1 + b_2)$.

5. What are the second and third steps in finding the area of the trapezoid?

Substitute 4 for *h*, 6 for b_1, and 8 for b_2. Add the lengths of the bases.

6. What is the area of the trapezoid?

28 in²

LESSON **10-2**

Puzzles, Twisters & Teasers

Hear, Hear!

Find the area of each figure below and circle the letter next to the correct answer. Write the letters over the number of the problem in the corresponding space in the riddle.

1.

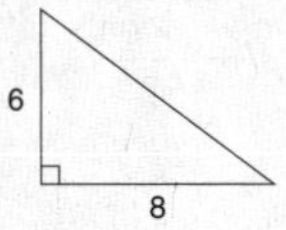

(E) 24 units^2 B 26 units^2 X 28 units^2

2.

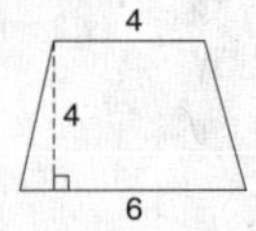

(A) 20 units^2 F 24 units^2 G 40 units^2

3.

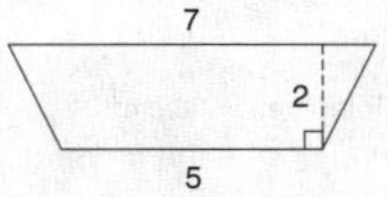

O 10 units^2 **(R)** 12 units^2 P 24 units^2

4.

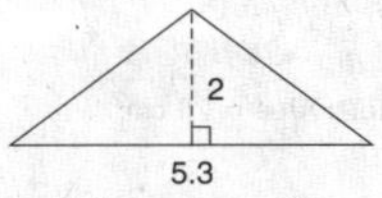

C 5.1 units^2 Q 5.2 units^2 **(D)** 5.3 units^2

5.

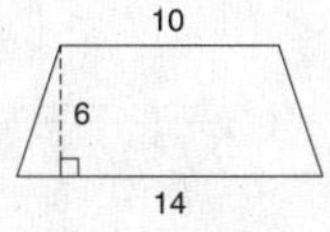

L 24 units^2 E 48 units^2 **(R)** 72 units^2

6.

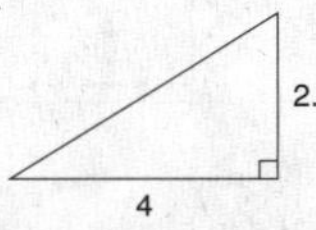

V 4 units^2 **(U)** 5 units^2 I 6 units^2

7.

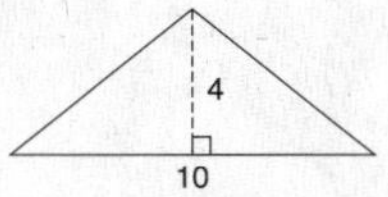

(M) 20 units^2 H 21 units^2 W 40 units^2

8.

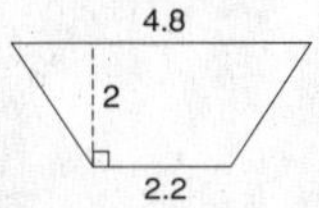

J 5 units^2 N 6 units^2 **(S)** 7 units^2

What part of your body has the most rhythm?

E	A	R	D	R	U	M	S
1.	2.	3.	4.	5.	6.	7.	8.

LESSON **10-3**

Practice A

Area of Composite Figures

Find the area of each polygon.

1.

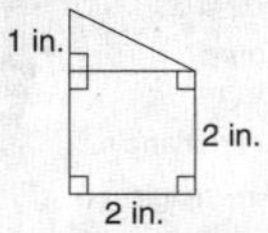

5 in^2

2.

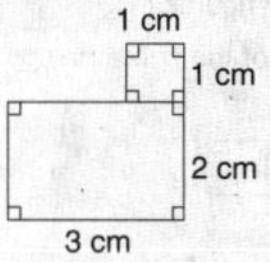

7 cm^2

3.

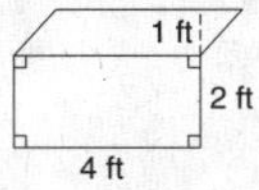

12 ft^2

4.

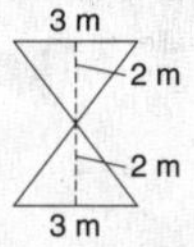

6 m^2

5.

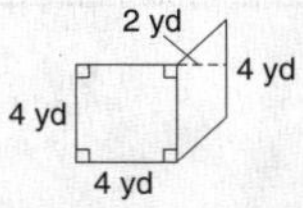

24 yd^2

6.

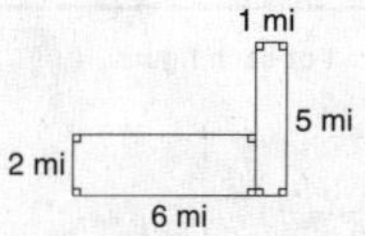

17 mi^2

7. A rectangular painting is made up of two congruent squares with sides that measure 2 feet. What is the area of the entire painting?

8 ft^2

8. A carpet is made up of two congruent triangles. The base of each triangle is 3 feet and the height is 6 feet. What is the area of the entire carpet?

18 ft^2

LESSON **10-3**

Practice B

Area of Composite Figures

Find the area of each polygon.

1.

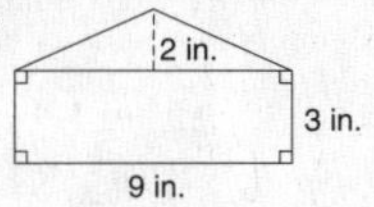

36 in^2

2.

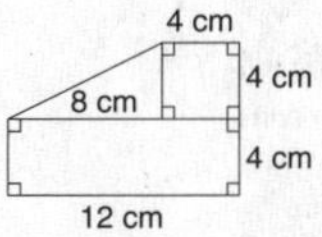

80 cm^2

3.

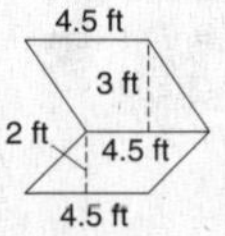

22.5 ft^2

4.

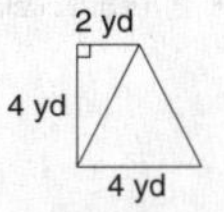

12 yd^2

5.

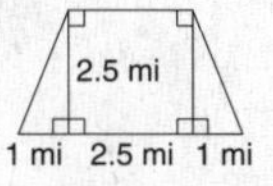

8.75 mi^2

6.

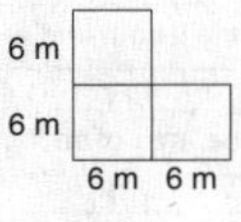

108 m^2

7. Three paintings are shaped like an 8-foot square, a 7-foot by 4-foot rectangle, and a triangle with a 6-foot base and a height of 7 feet. If those paintings are hung together on the outside of a building, how much of the building's wall will they cover altogether?

113 ft^2

8. Two diagonals divide a square carpet into 4 congruent triangles. The base of each triangle is 5 feet and the height is 2.5 feet. What is the area of the entire carpet?

25 ft^2

LESSON **10-3**

Practice C

Area of Composite Figures

Find the area of each figure. The shaded parts are cut out of the figures.

1.

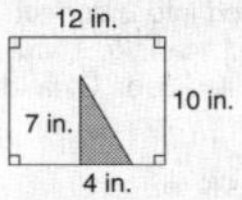

106 in^2

2.

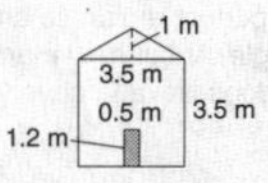

13.4 m^2

3.

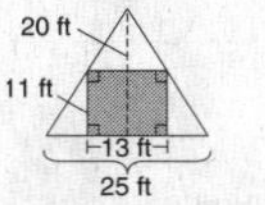

107 ft^2

4.

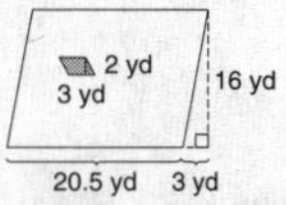

322 yd^2

5.

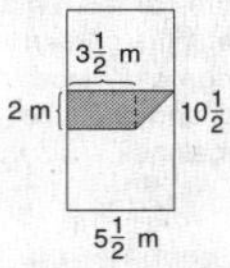

$48\frac{3}{4}$ m^2

6.

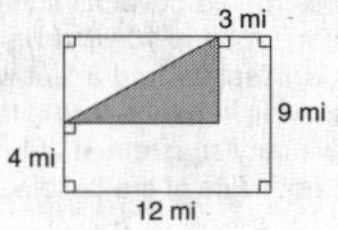

85.5 mi^2

7. The front wall of the shed that Rita built is 9 feet tall and $5\frac{1}{4}$ feet wide. After she cut a section out for a door, the area of that wall was $26\frac{1}{4}$ square feet. If the measurements of the doorway were whole numbers of feet, what size door did Rita install?

a door that is 7 feet tall and 3 feet wide

8. The perimeter of a rectangular playground is 40 yards. It is $1\frac{1}{2}$ yards longer than it is wide. What is the area of that playground?

$99\frac{7}{16}$ yd^2

LESSON 10-3
Reteach
Area of Composite Figures

Sometimes you can use area formulas you know to help you find the area of other figures.

To find the area of the figure below, first divide the figure into figures you know.

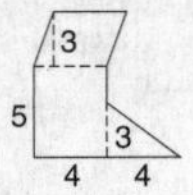

The figure is made up of a triangle, a parallelogram, and a rectangle.

Next, find the area of each figure.

Triangle	Parallelogram	Rectangle
$A = \frac{1}{2}bh$	$A = bh$	$A = \ell w$
$= \frac{1}{2}(3 \cdot 4)$	$= 3 \cdot 4$	$= 4 \cdot 5$
$= 6$	$= 12$	$= 20$

Then, find the sum of all of the areas.

$6 + 12 + 20 = 38$ The area of the figure is 38 square units.

Find the area of each figure.

1. 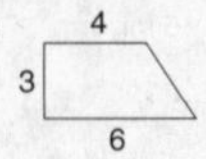

15 square units

2. 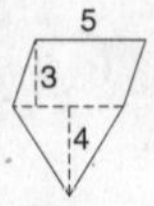

25 square units

3. 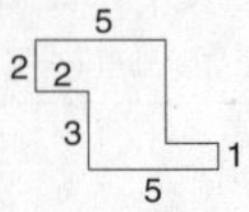

21 square units

4. 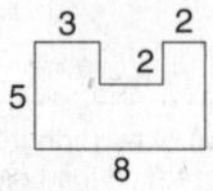

34 square units

LESSON 10-3
Challenge
Area Combinations

Combine these polygons to draw figures with the given areas. You must use all the polygons, and each can be used only once. **Sample answers given.**

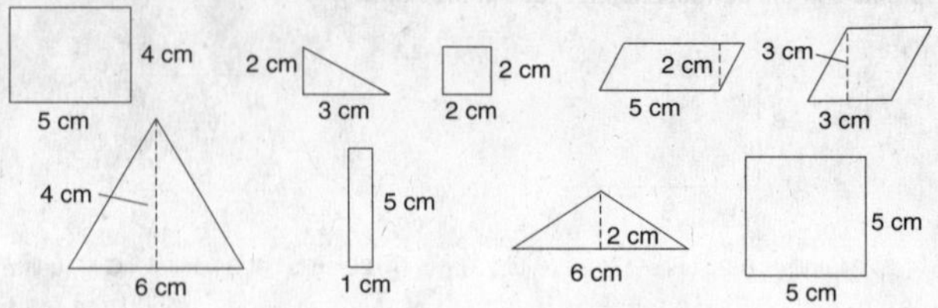

1. Total Area = 26 cm^2

2. Total Area = 30 cm^2

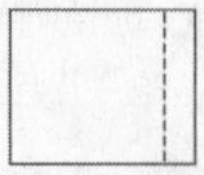

3. Total Area = 22 cm^2

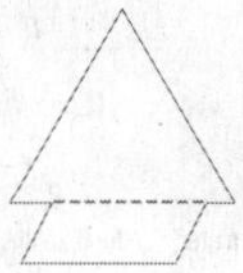

4. Total Area = 16 cm^2

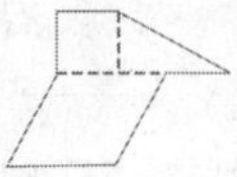

LESSON 10-3
Problem Solving
Area of Composite Figures

Write the correct answer.

1. The shape of Nevada can almost be divided into a perfect rectangle and a perfect triangle. About how many square miles does Nevada cover?

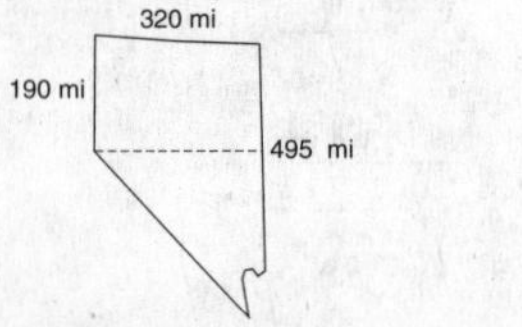

about 109,600 mi^2

2. The shape of Oklahoma can almost be divided into 2 perfect rectangles and 1 triangle. About how many square miles does Oklahoma cover?

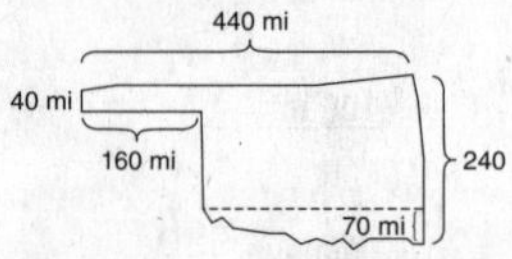

about 63,800 mi^2

3. The front side of an apartment building is a rectangle 60 feet tall and 25 feet wide. Bricks cover its surface, except for a door and 10 windows. The door is 7 feet tall and 3 feet wide. Each window is 4 feet tall and 2 feet wide. How many square feet of bricks cover the front side of the building?

1,399 ft^2

4. Each side of a square garden is 12 meters long. A hedge wall 1 meter wide surrounds the garden. What is the area of the entire garden including the hedge wall? How many square meters of land does the hedge wall cover alone?

entire garden = 196 m^2; hedge wall = 52 m^2

Circle the letter of the correct answer.

5. A figure is formed by a square and a triangle. Its total area is 32.5 m^2. The area of the triangle is 7.5 m^2. What is the length of each side of the square?

(A) 5 meters C 15 meters
B 25 meters D 16.25 meters

6. A rectangle is formed by two congruent right triangles. The area of each triangle is 6 in^2. If each side of the rectangle is a whole number of inches, which of these could not be its perimeter?

F 26 inches (H) 24 inches
G 16 inches J 14 inches

LESSON 10-3
Reading Strategies
Use a Flowchart

To find the area of a rectangle, multiply the length by the width.

$A = \ell \cdot w$

$A = 4 \text{ units} \cdot 5 \text{ units}$

$A = 20 \text{ square units}$

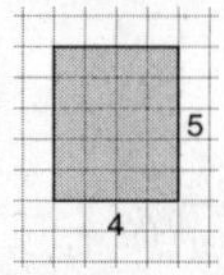

If a polygon is made up of more than one rectangle, you can find its area by using the flowchart below.

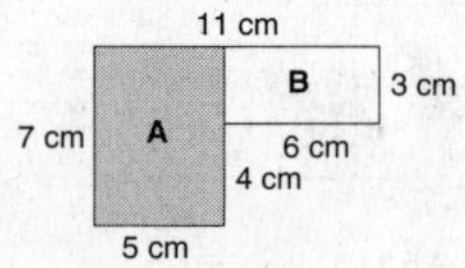

Step 1: Separate the polygon into rectangles.
↓
Step 2: Mark the length and width of each rectangle.
↓
Step 3: Find the area of each rectangle.
↓
Step 4: Add the areas of all of the rectangles together.

Answer the following questions.

1. What is the formula for the area of a rectangle? $A = \ell \cdot w$
2. What are the length and width of rectangle A? length: 7 cm; width: 5 cm
3. What is the area of rectangle A? 35 cm^2
4. What is the area of rectangle B? 18 cm^2
5. What do you do with the areas of rectangles A and B to find the area of the polygon? add them
6. What is the area of the polygon? 53 cm^2

LESSON **10-3** **Puzzles, Twisters & Teasers**
Sky High

Find the area of each figure. Circle the answers at the bottom of the page. Rearrange the circled letters to solve the riddle.

1.
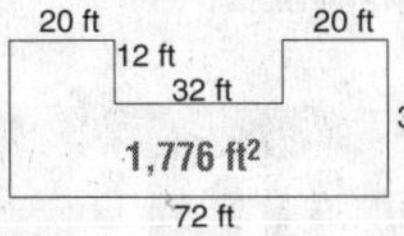

2.
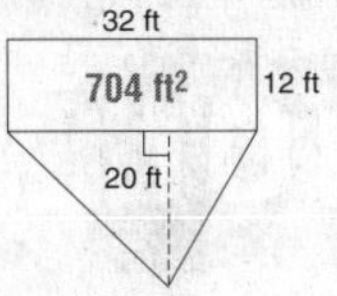

3.
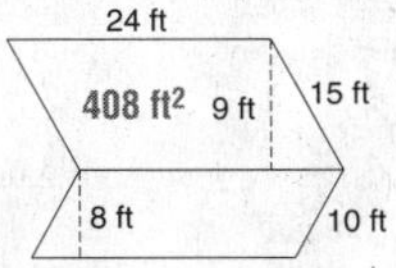

4.
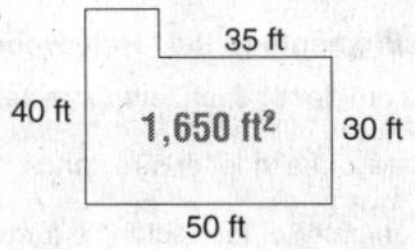

5.
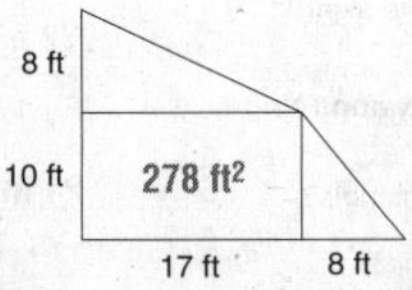

6.
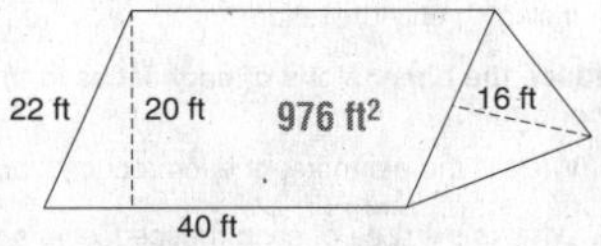

What bird can lift the heaviest weights? a crane

S 724 ft²	Ⓐ 408 ft²	W 386 ft²	Ⓔ 1,776 ft²
Ⓐ 704 ft²	L 880 ft²	Ⓝ 976 ft²	I 1,232 ft²
M 440 ft²	Ⓒ 278 ft²	T 1,024 ft²	Ⓡ 1,650 ft²

LESSON **10-4** **Practice A**
Comparing Perimeter and Area

Write how the perimeter and the area of the figure change when its dimensions change.

1.

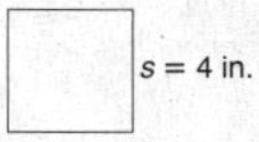

$P =$ 8 in. $P =$ 16 in.

$A =$ 4 in² $A =$ 16 in²

When the dimensions of the square are doubled, the perimeter is doubled, and the area is 4 times greater.

2. Use a centimeter ruler to measure the square. Then draw another square with dimensions that are half as great as the given square. How do the perimeter and the area change when the dimensions change?

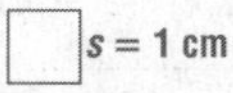

When the dimensions of the square are half as great, the perimeter is divided by 2, and the area is divided by 4.

3. A rectangular photo is 5 inches long and 2 inches wide. Jin-Sook wants to enlarge the photo by doubling its length and width. How many inches of wood will she need to make a frame for the enlarged photo?

28 inches of wood

4. How much glass will Jin-Sook need to cover the enlarged photo from Exercise 3?

40 square inches of glass

LESSON **10-4** **Practice B**
Comparing Perimeter and Area

Write how the perimeter and the area of the figure change when its dimensions change.

1.
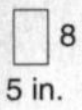

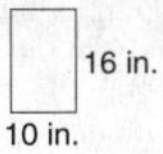

$P =$ 26 in. $P =$ 52 in.

$A =$ 40 in² $A =$ 160 in²

When the dimensions of the rectangle are doubled, the perimeter is doubled, and the area is 4 times greater.

2. Use a centimeter ruler to measure the triangle. Then draw another triangle with dimensions that are half as great as the given triangle. How do the perimeter and the area change when the dimensions change?

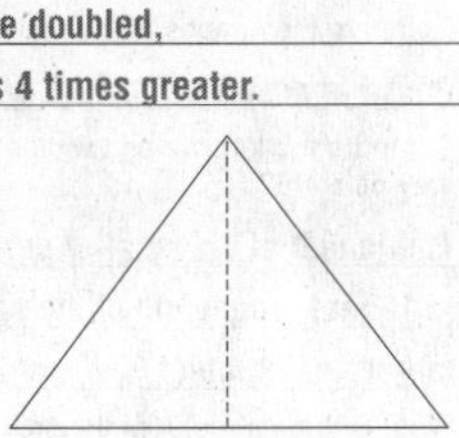

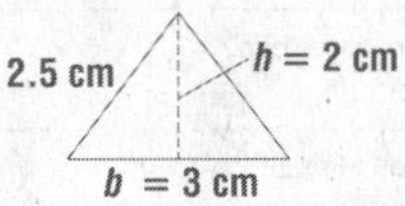

When the dimensions of the triangle are half as great, the perimeter is divided by 2, and the area is divided by 4.

3. Nina wants to make a smaller version of a painting she saw in a museum. The museum painting was a square with each side measuring 6.4 feet. If Nina makes her copy half the size of the original painting, how much space will it cover on her wall?

10.24 square feet

4. How many feet of wood will Nina need to make a frame for her painting from Exercise 3?

12.8 feet of wood

LESSON **10-4** **Practice C**
Comparing Perimeter and Area

Write how the perimeter and the area of the figure change when its dimensions change.

1.
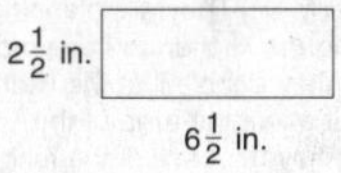

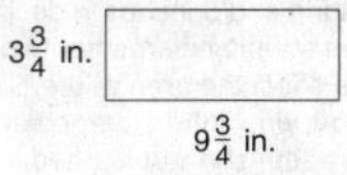

$P =$ 18 in. $P =$ 27 in.

$A =$ $16\frac{1}{4}$ in² $A =$ $36\frac{9}{16}$ in²

When the dimensions of the rectangle are $1\frac{1}{2}$ times greater, the perimeter is $1\frac{1}{2}$ times greater, and the area is $2\frac{1}{4}$ times greater.

2. Use a centimeter ruler to measure the rectangle. Then draw another rectangle with dimensions that are one-half as great as the given rectangle. How do the perimeter and the area change when the dimensions change?

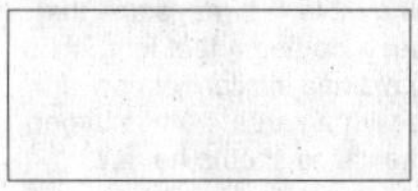

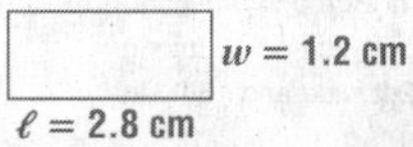

When the dimensions of the rectangle are half as great, the perimeter is divided by 2, and the area is divided by 4.

3. The area of a rectangular photo is 80 square inches. After reducing its dimensions, the area of the photo is 5 square inches. By how much were the dimensions of the original photo reduced?

by one-fourth

4. The side lengths of the original and reduced photo in Exercise 3 are whole numbers. If the length of the original photo is 10 inches, what is the perimeter of the reduced photo?

9 inches

LESSON **10-4**

Reteach
Comparing Perimeter and Area

When the dimensions of a figure change, the perimeter and area of the figure change.

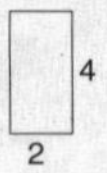

$P = 2\ell + 2w$ $\qquad A = \ell w$

$= (2 \cdot 4) + (2 \cdot 2)$ $\qquad = 4 \cdot 2$
$= 8 + 4$ $\qquad = 8$ square units
$= 12$ units

The dimensions are $\ell = 4$ and $w = 2$.

If each dimension of the rectangle is divided by 2, the perimeter and area change.

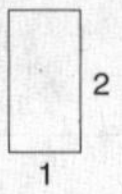

$P = 2\ell + 2w$ $\qquad A = \ell w$
$= (2 \cdot 2) + (2 \cdot 1)$ $\qquad = 2 \cdot 1$
$= 4 + 2$ $\qquad = 2$ square units
$= 6$ units

The new dimensions are $\ell = 2$ and $w = 1$.

When the dimensions of a rectangle are divided by 2, the perimeter is divided by 2 and the area is divided by 4.

Write how the perimeter and area change when the dimensions change.

1.

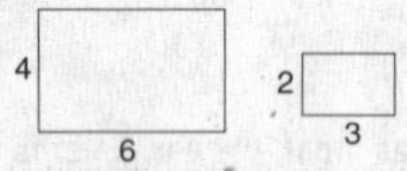

When the dimensions of a rectangle are divided by 2, the perimeter is divided by 2 and the area is divided by 4.

2.

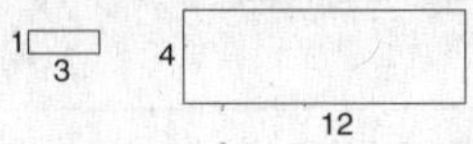

When the dimensions of a rectangle are multiplied by 4, the perimeter is multiplied by 4 and the area is multiplied by 16.

LESSON **10-4**

Challenge
Hooray for Hollywood!

The "Hollywood" sign in Los Angeles, California, is one of the most recognized landmarks of the United States. Each of the sign's letters covers a rectangular space 9 meters wide and 12 meters tall.

Built in 1923, the original sign said "Hollywoodland."

Find these measurements of the "Hollywood" sign.

1. What is the perimeter of each letter's rectangle? **42 m**
2. What is the area of each letter's rectangle? **108 m^2**
3. What is the perimeter of the rectangle formed by the entire sign, including only the letters? **186 meters**
4. What is the area the rectangle formed by the entire sign, including only the letters? **972 m^2**

Reduce the dimensions of each letter in the "Hollywood" sign by one-half.

5. What is the perimeter of each reduced letter's rectangle? **21 m**
6. What is the area of each reduced letter's rectangle? **27 m^2**
7. What is the perimeter of the rectangle formed by the entire reduced sign, including only the letters? **93 m**
8. What is the area of the rectangle formed by the entire reduced sign, including only the letters? **243 m^2**
9. How does the perimeter of each letter change when its dimensions change?
   When each letter is reduced by one-half, its perimeter is divided by 2.
10. How does the area of the entire sign change when its dimensions change?
   When the entire sign is reduced by one-half, its area is divided by 4.

LESSON **10-4**

Problem Solving
Comparing Perimeter and Area

Write the correct answer.

1. Fiona's school photograph is 6 inches long and 5 inches wide. If she orders a triple enlargement how would this affect the area of the photo? How would the enlargement affect the frame she would need for the photo?

   The area of the enlarged photo will be 9 times larger than the original. She will need a frame 3 times larger than the frame for the original photo.

2. The Whitman's kitchen is 8 feet long and 6 feet wide. They are planning on renovating the kitchen to have more space. If they double just the width, how will it affect the area of the room? If they double just the length? If they double both measurements?

   If they double just 1 measurement, the area would double. If they double both measurements, the room's area would be 4 times larger.

Circle the letter of the correct answer.

3. Kent saw a table in a magazine that was 3 feet wide and 4 feet long. If he wants to make a similar version of the table with an area 4 times larger, what dimensions should he use?
   - **A** 4 ft wide and 5 ft long
   - **(B)** 6 ft wide and 8 ft long
   - **C** 9 ft wide and 12 ft long
   - **D** 12 ft wide and 16 ft long

4. The triangular sail on Shakeera's boat is 8 meters wide and 10 meters tall. She wants to make a model of the boat that is $\frac{1}{20}$ of its actual size. How much canvas will Shakeera use for the model boat's sail?
   - **F** 10 m^2 of canvas
   - **G** 1 m^2 of canvas
   - **(H)** 0.1 m^2 of canvas
   - **J** 0.01 m^2 of canvas

5. A triangle is 6.4 cm long and 8.2 cm tall. If you triple its dimensions, what would be the area of the enlarged triangle?
   - **A** 78.72 cm^2
   - **(C)** 236.16 cm^2
   - **B** 157.44 cm^2
   - **D** 472.32 cm^2

6. The dimensions of a regular pentagon are doubled. The perimeter of the enlarged pentagon is 25 yards. What was the length of each side of the original pentagon?
   - **(F)** 2.5 yards
   - **H** 5 yards
   - **G** 12 yards
   - **J** 16.25 yards

LESSON **10-4**

Reading Strategies
Compare and Contrast

Mark is trying to decide which of two different pens to buy for his dog.

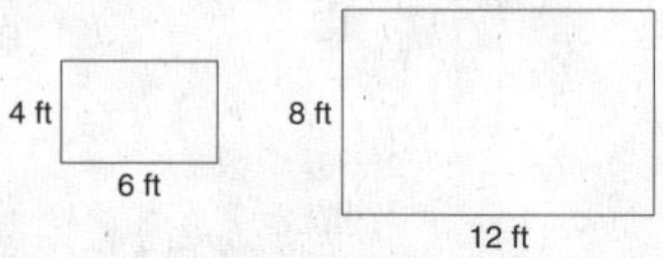

Answer the following questions.

1. Compare the shapes of the two pens. How are they alike?
   Possible answer: They are both rectangles.
2. Compare the lengths and widths of the two pens. How are they different?
   The length of the smaller pen is 6 feet, and the length of the larger pen is 12 feet. The width of the smaller pen is 4 feet, and the width of the larger pen is 8 feet.
3. What is the perimeter of the smaller pen? Of the larger pen?
   20 feet; 40 feet
4. How do the perimeters of the two pens compare in length?
   The perimeter of the larger pen is twice as long as the perimeter of the smaller pen.
5. What is the area of the smaller pen? Of the larger pen?
   24 feet^2; 96 feet^2
6. How do the areas of the two pens compare in size?
   The area of the larger pen is four times the area of the smaller pen.

LESSON **10-4**

Puzzles, Twisters & Teasers

Truth or Dare?

There are three correct answers for each problem. Circle the incorrect statement. The letters next to the incorrect statements will provide the solution to the riddle.

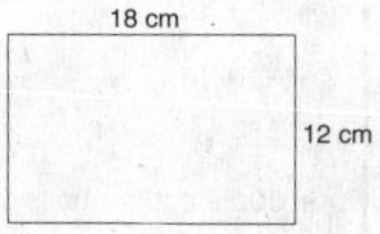

1. **B** If you double the dimensions, the new perimeter will be 120 cm.
   F The perimeter of the figure is 60 cm.
   (R) If you double the dimensions, the new area will be 432 cm^2.
   T If you halve the dimensions, the new perimeter will be 30 cm.

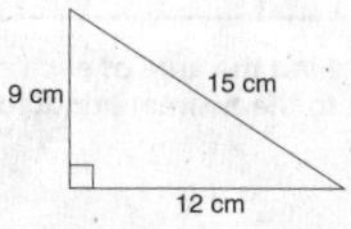

2. **(A)** If you triple the dimensions, the new perimeter will be 324 cm.
   C If you divide the dimensions in half, the new perimeter will be 18 cm.
   I If you triple the dimensions, the new area will be 9 times as large.
   S The area of the figure is 54 cm^2.

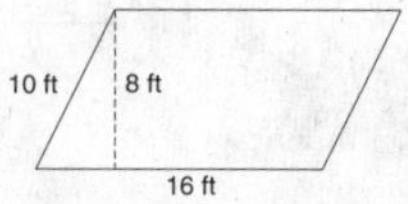

3. **L** The area of the figure is 128 ft^2.
   W The perimeter of the figure is 52 ft.
   M If you halve the dimensions, the area will be 32 ft^2.
   (A) If you double the dimensions, the new perimeter will be four times as large.

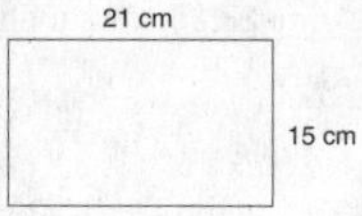

4. **W** If you multiply the dimensions by 4, the new perimeter will be 288 cm.
   (S) The area of the figure is 157.5 cm^2.
   E If you divide the dimensions by 3, the new area will be 35 cm^2.
   G The perimeter of the figure is 72 cm.

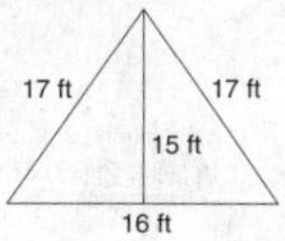

5. **P** The area of the figure is 120 ft^2.
   H The perimeter of the figure is 50 ft.
   C If you multiply the dimensions by 5, the new perimeter will be 5 times as large.
   (T) If you multiply the dimensions by 3, the new area will be 6 times as large.

What is higher than a house and seems smaller than a mouse?

A		S	T	A	R
3		4	5	3	1

LESSON **10-5**

Practice A

Area of Circles

Estimate the area of each circle.

1. 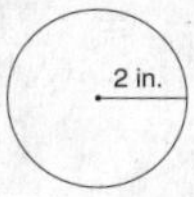

$A \approx$ **12 $in.^2$**

2. 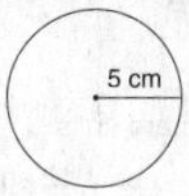

$A \approx$ **75 cm^2**

3. 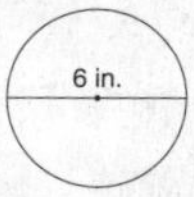

$A \approx$ **27 $in.^2$**

4. 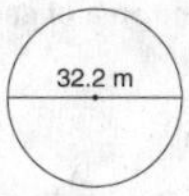

$A \approx$ **768 m^2**

Find the area of each circle. Use $\frac{22}{7}$ for pi.

5. 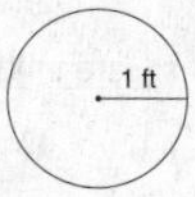

$A \approx$ **$\frac{22}{7}$ ft^2**

6. 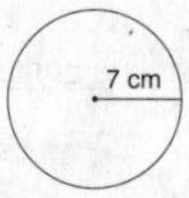

$A \approx$ **154 cm^2**

7. 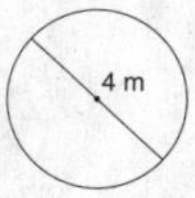

$A \approx$ **12.57 m^2**

8. 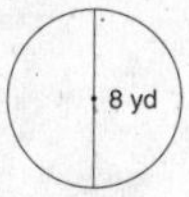

$A \approx$ **50.28 yd^2**

9. The Mayall Telescope at Kitt Peak National Observatory in Arizona has a lens with a 2-meter radius. This glass lens is covered with a thin sheet of reflective aluminum. How many square meters of glass does the aluminum cover? Use 3.14 for pi.

12.56 square meters

LESSON **10-5**

Practice B

Area of Circles

Estimate the area of each circle.

1. 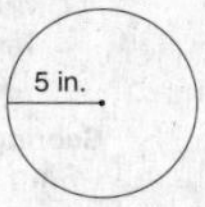

$A \approx$ **75 in^2**

2. 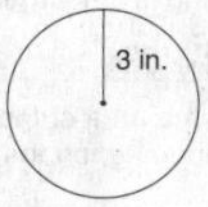

$A \approx$ **27 in^2**

3. 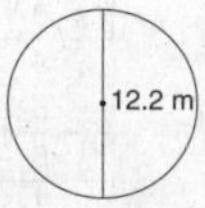

$A \approx$ **108 m^2**

4. 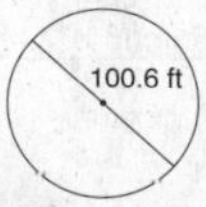

$A \approx$ **7,500 ft^2**

Find the area of each circle. Use $\frac{22}{7}$ for *pi*.

5. 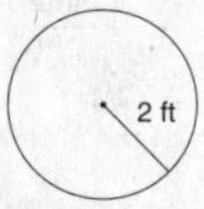

$A \approx$ **12.57 ft^2**

6. 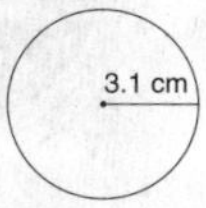

$A \approx$ **30.2 cm^2**

7. 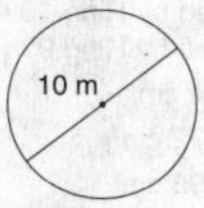

$A \approx$ **78.57 m^2**

8. 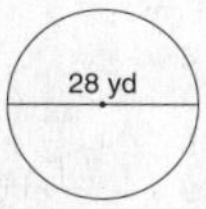

$A \approx$ **616 yd^2**

9. Stonehenge, a circle of large carved stones in England, was built more than 1,000 years ago. The circle of stones has a diameter of 108 feet. About how many square feet of land does Stonehenge cover? Use $\frac{22}{7}$ for *pi*.

9,164.57 square feet

LESSON **10-5**

Practice C

Area of Circles

Estimate the area of each circle.

1. 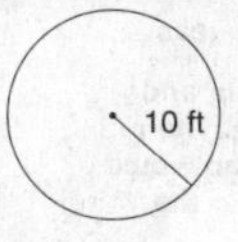

$A \approx$ **300 ft^2**

2. 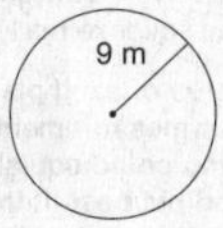

$A \approx$ **243 m^2**

3. 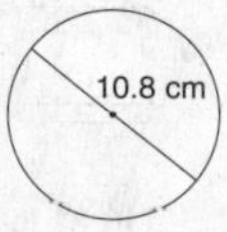

$A \approx$ **75 cm^2**

4. 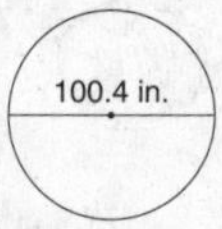

$A \approx$ **7,500 in^2**

Find the area of each circle. Use $\frac{22}{7}$ for *pi*.

5. 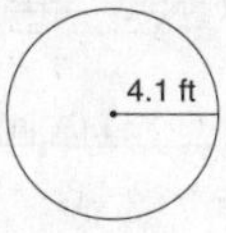

$A \approx$ **52.83 ft^2**

6. 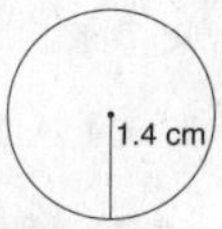

$A \approx$ **6.16 cm^2**

7. 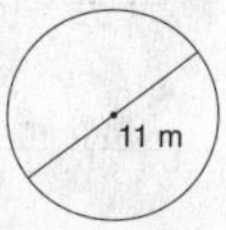

$A \approx$ **95.07 m^2**

8. 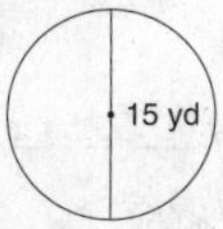

$A \approx$ **176.79 yd^2**

9. The diameter of a sandwich cookie is 1.75 inches. To the nearest tenth, what is the combined area of the two wafer parts of the sandwich cookie? Use 3.14 for *pi*.

4.8 square inches

LESSON **10-5**

Reteach

Area of Circles

To estimate the area A of a circle, use the formula $A = \pi r^2$ and 3 for π.

$A = \pi r^2$
$\approx 3 \cdot 2^2$
$\approx 3 \cdot 4$
≈ 12 square units.

The area is about 12 square units.

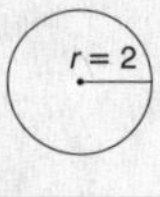

Estimate the area of each circle. Use 3 for π.

1.

$A \approx$ 3 square units

2.

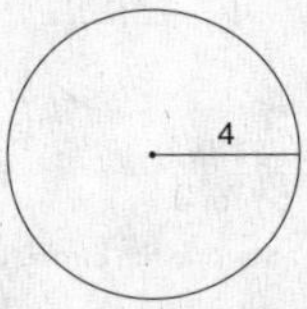

$A \approx$ 48 square units

3.

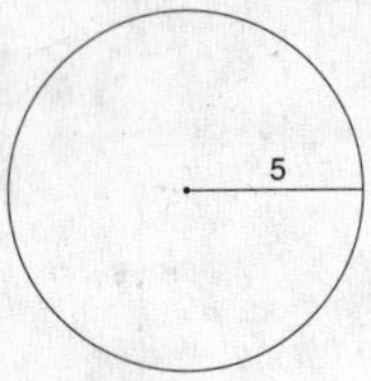

$A \approx$ 75 square units

4.

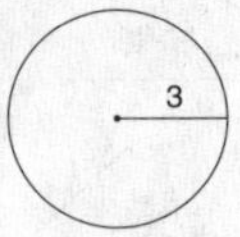

$A \approx$ 27 square units

LESSON **10-5**

Reteach

Area of Circles (continued)

To find the area A of a circle, use the formula $A = \pi r^2$ and $\frac{22}{7}$ for π.

$A = \pi r^2$
$\approx \frac{22}{7} \cdot 4^2$
$\approx \frac{22}{7} \cdot 16$
$\approx \frac{352}{7}$
≈ 50.29 square units

The area is about 50.29 square units.

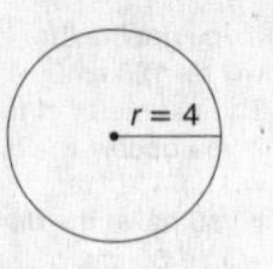

Find the area of each circle. Use $\frac{22}{7}$ for π. Round each answer to the nearest hundredth.

5.

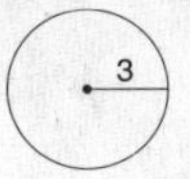

$A \approx$ 28.29 square units

6.

$A \approx$ 3.14 square units

7.

$A \approx$ 12.57 square units

8.

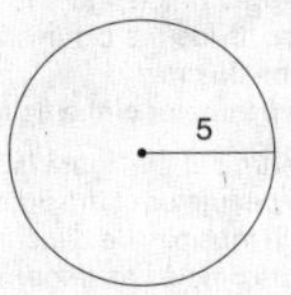

$A \approx$ 78.57 square units

LESSON **10-5**

Challenge

Pies Are Squared

Have you ever heard the saying, "You can't fit a square peg in a round hole"? Well, sometimes you can fit a round pie in a square dish—if you squeeze really hard and don't mind messy areas.

Find the area of each pie and pie dish. Use 3.14 for π, and round your measurements to the nearest hundredth. Then imagine you could squish the pies. Which square pan would each round pie best fit in? Draw a line from the pie to the appropriate pie dish.

1.

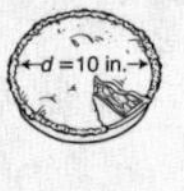

Area = 78.5 in^2

A.

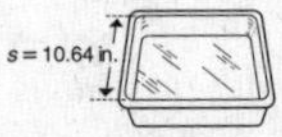

Area = 113.21 in^2

2.

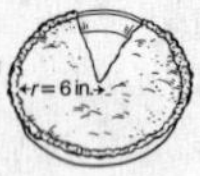

Area = 113.04 in^2

B.

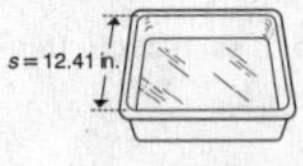

Area = 154.01 in^2

3.

Area = 63.59 in^2

C.

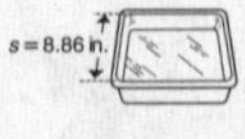

Area = 78.5 in^2

4.

Area = 153.86 in^2

D.

Area = 63:68 in^2

LESSON **10-5**

Problem Solving

Area of Circles

Use the table to answer each question. Use 3.14 for *pi*.

Official Archery Target Ring Diameters

Scoring Ring	Diameter (cm)
1	80
2	72
3	64
4	56
5	48
6	40
7	32
8	24
9	16
10	8
Inner 10	4

1. Which ring is the largest? What area does it enclose?

   Ring 1; 5,024 cm^2

2. What is the area of the center circle, or the inner 10 scoring ring, on the target?

   12.56 cm^2

3. What area does Ring 5 enclose?

   1,808.64 cm^2

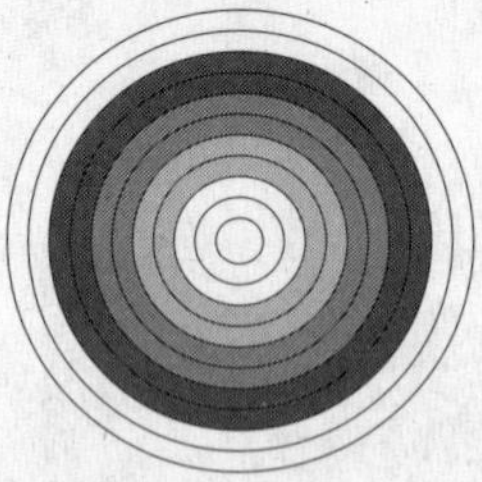

Circle the letter of the correct answer.

4. Which ring encloses an area of 4069.44 cm^2?
   - (A) Ring 2
   - B Ring 3
   - C Ring 6
   - D Ring 8

5. How much greater is the area enclosed by Ring 10 than the area enclosed by Ring 9?
   - F 50.24 cm^2
   - (G) 150.72 cm^2
   - H 200.96 cm^2
   - J 251.2 cm^2

6. What is the area enclosed by Ring 6?
   - A 5,024 cm^2
   - B 1,600 cm^2
   - (C) 1,256 cm^2
   - D 62.8 cm^2

7. What is the area enclosed by Ring 1?
   - F 10 times the area of Ring 10
   - G 20 times the area of Ring 10
   - (H) 100 times the area of Ring 10
   - J 1,000 times the area of Ring 10

LESSON **10-5**
Reading Strategies
Focus on Vocabulary

The **area** of a circle is equal to *pi* times the radius squared.

Pi is represented by the symbol π. Use $\frac{22}{7}$ or 3.14 for *pi*.

The **formula** for finding the area of a circle is $A = \pi r^2$.

You can **estimate** the area of a circle by using 3 to approximate the value of *pi*.

Answer each question.

1. What is the formula for finding the area of a circle?
   $A = \pi r^2$
2. How do you estimate the area of a circle?
   Substitute 3 for *pi* in the formula for finding the area of a circle.
3. What is the symbol for *pi*?
   π
4. What words can you use to tell the formula for finding the area of a circle?
   The area of a circle is equal to *pi* times the radius squared.
5. A circle has a radius of 5 inches. Estimate the area.
   about 75 square inches
6. Use 3.14 to find the area of the circle in Problem 5.
   78.5 square inches
7. A circle has a radius of 7 centimeters. Estimate the area.
   about 147 square centimeters
8. Use $\frac{22}{7}$ to find the area of the circle in Problem 7.
   154 square centimeters

LESSON **10-5**
Puzzles, Twisters & Teasers
You're All Wet

Find the area of each circle below. Use 3.14 for *pi*. To solve the riddle, match the letter of the area to the problem number.

Area (square units)	Letter
63.59	O
65.39	I
94.99	M
99.49	N
30.18	T
31.08	S
132.67	T
123.67	R
19.63	O
131.04	E
113.04	B

1. 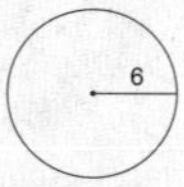

113.04

2. 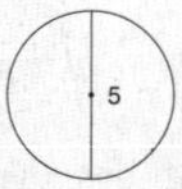

19.63

3. 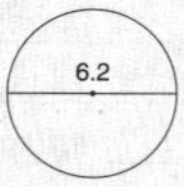

30.18

4. 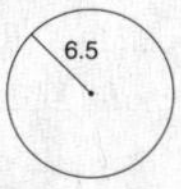

132.67

5. 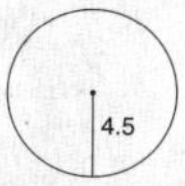

63.59

6. 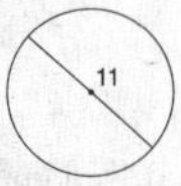

94.99

Where is the ocean the deepest?

ON THE B O T T O M
1. 2. 3. 4. 5. 6.

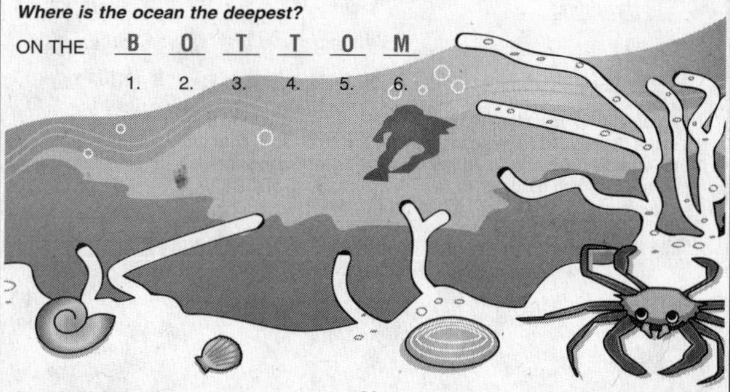

LESSON **10-6**
Practice A
Three-Dimensional Figures

Use the polyhedron to answer each question.

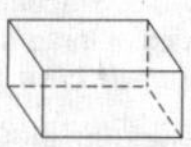

1. How many faces does it have? 6
2. How many edges does it have? 12
3. How many vertices does it have? 8
4. What is the shape of its bases?
   rectangle
5. What kind of polyhedron is it?
   rectangular prism

Tell whether each figure is a polyhedron and name the three-dimensional figure.

6. 
no; a cylinder

7. 
yes; triangular prism

8. 
yes; rectangular pyramid

9. LeJuan bought a book for his sister's birthday. What shape box should LeJuan put the book in before wrapping it?
   a rectangular prism
10. Look around your classroom. Find 5 three-dimensional figures and name what kind each is.
   Answers will vary, but should identify and name 5 three-dimensional figures in the classroom. Possible answers include eraser: rectangular prism; chalk: cylinder; globe: sphere; mug: cylinder; desk: rectangular prism.

LESSON **10-6**
Practice B
Three-Dimensional Figures

Identify the number of faces, edges, and vertices in each three-dimensional figure.

1. 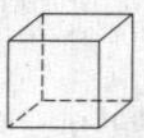
6 faces; 12 edges; 8 vertices

2. 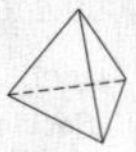
4 faces 6 edges; 4 vertices

3. 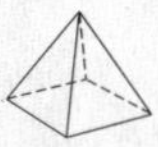
5 faces; 8 edges; 5 vertices

Tell whether each figure is a polyhedron and name the three-dimensional figure.

4. 
no; cone

5. 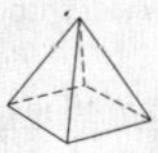
yes; rectangular pyramid

6. 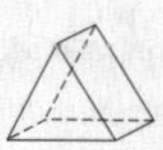
yes; triangular prism

7. Kelly wants to make a box in the shape of a cube. How many pieces of wood does she need? In what shape should she cut them? Explain.
   She needs 6 square pieces of wood because a cube has 6 square faces.
8. Kwan made a sculpture in the shape of a polyhedron. It only has one base that is a triangle. What three-dimensional figure is her sculpture? Explain your reasoning.
   It is a triangular pyramid, because a pyramid has only 1 base, and the shape of that base defines what kind of pyramid it is.

LESSON **10-6**

Practice C

Three-Dimensional Figures

Tell whether each figure is a polyhedron and name the three-dimensional figure. Then identify the number of faces, edges, and vertices in each three-dimensional figure.

1. 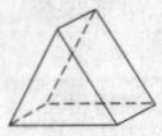

yes; triangular prism; 5 faces; 9 edges; 6 vertices

2. 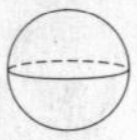

no; sphere; no faces; no edges; no vertices

3. 

yes; hexagonal pyramid; 7 faces; 12 edges; 7 vertices

4. 

no; cylinder; 2 faces; no edges; no vertices

5. 

yes; pentagonal prism; 7 faces; 15 edges; 10 vertices

6. 

yes; octagonal pyramid; 9 faces; 16 edges; 9 vertices

7. A construction company is building two hotels that will be the same height, and have bases that are the same size. One will be a rectangular prism and the other will be a rectangular pyramid. Which building will require more gallons of paint to completely cover it? Why? **Possible answer:**

the rectangular prism building, because it has one more side to paint.

LESSON **10-6**

Reteach

Three-Dimensional Figures

A polyhedron is a three-dimensional figure. A polyhedron has flat surfaces, called faces. Faces that share a side form an edge. The point at which three or more edges meet is called a vertex.

This polyhedron has 4 faces, 4 vertices, and 6 edges.

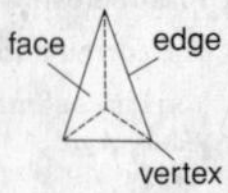

Identify the number of faces, edges, and vertices on each three-dimensional figure.

1. 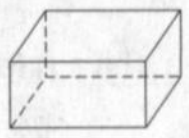

6, 12, 8

2. 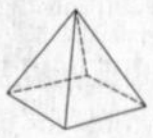

5, 8, 5

A prism is a three-dimensional figure with two congruent parallel bases. The other faces are all parallelograms.

A pyramid has one base. The other faces are triangles.

A cylinder has two circular parallel bases joined by a curved surface.

A cone has one circular base and a curved surface that comes to a point.

Pyramids and prisms are polyhedrons because each face is a polygon. Cylinders and cones are not polyhedrons because they have curved surfaces.

Tell whether each figure is a polyhedron and name each three-dimensional figure.

3. 

no, cone

4. 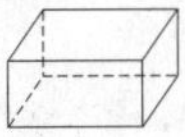

yes, rectangular prism

LESSON **10-6**

Challenge

Polyhedron Patterns

Complete these charts to discover the polyhedron patterns.

	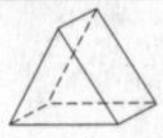 Triangular Prism	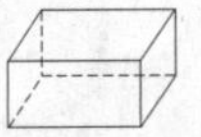 Rectangular Prism	 Pentagonal Prism	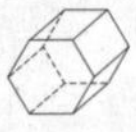 Hexagonal Prism
Base's Number of Sides	3	4	5	6
Faces	5	6	7	8
Vertices	6	8	10	12
Edges	9	12	15	18

PRISM PATTERNS: If n = the number of sides on the base of a prism, what three expressions show that prism's number of faces, vertices, and edges?

faces $= n + 2$; vertices $= 2n$; edges $= 3n$

	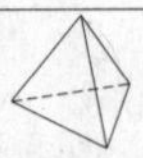 Triangular Pyramid	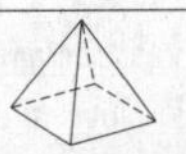 Rectangular Pyramid	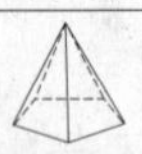 Pentagonal Pyramid	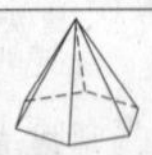 Hexagonal Pyramid
Base's Number of Sides	3	4	5	6
Faces	4	5	6	7
Vertices	4	5	6	7
Edges	6	8	10	12

PYRAMID PATTERNS: If n = the number of sides on the base of a pyramid, what three expressions show that pyramid's number of faces, vertices, and edges?

faces $= n + 1$; vertices $= n + 1$; edges $= 2n$

LESSON **10-6**

Problem Solving

Three-Dimensional Figures

Write the correct answer.

1. Pamela folded an origami figure that has 5 faces, 8 edges, and 5 vertices. What kind of three-dimensional figure could Pamela have created?

   a rectangular or square pyramid

2. Look at your classroom chalkboard. What kind of three-dimensional figure is the board eraser? What kind of three-dimensional figure is the chalk?

   eraser: rectangular prism; chalk: cylinder

3. If you cut a cylinder in half between its two bases, what two three-dimensional figures are formed?

   2 cylinders

4. You have two hexagons. How many rectangles do you need to create a hexagonal prism?

   6 rectangles

5. All four of the faces of a paperweight are triangles. Is this enough information to classify this three-dimensional figure? Explain.

   Yes, It is a triangular pyramid.

6. Paulo says that if you know the number of faces a pyramid has, you also know how many vertices it has. Do you agree? Explain.

   Yes; A pyramid always has the same number of faces and vertices.

Circle the letter of the correct answer.

7. How is a triangular prism different from a triangular pyramid?
   - **(A)** The prism has 2 bases.
   - B The pyramid has 2 bases.
   - C All of the prism's faces are triangles.
   - D The pyramid has 5 faces.

8. Which of these statements is not true about a cylinder?
   - F It has 2 circular bases.
   - G It has a curved lateral surface.
   - H It is a solid figure.
   - **(J)** It is a polyhedron.

9. A museum needs to ship a sculpture that has a curved lateral surface and one flat circular base. In what shape box should they mail the sculpture?
   - **(A)** cone
   - B cube
   - C cylinder
   - D triangular prism

10. A glass prism reflects white light as a multicolored band of light called a spectrum. The prism has 5 glass faces with 9 edges and 6 vertices. What kind of prism it it?
    - F cube
    - G cone
    - H triangular pyramid
    - **(J)** triangular prism

LESSON 10-6
Reading Strategies
Use a Graphic Organizer

This graphic organizer will help you learn about three-dimensional figures called **polyhedrons.**

Definition
A three-dimensional geometric figure that has four or more flat faces.

Prism
- Has two congruent parallel bases.
- All faces are polygons.
- Named by shape of its base.

Polyhedrons

Pyramid
- Has one polygon-shaped base.
- Other faces are triangles.
- Triangular faces meet at a common vertex.
- Named by the shape of its base.

Non-examples
Cylinder and cone
- Bases are circular.
- Surface is curved.

Use the graphic organizer to answer each question.

1. What are the flat surfaces of a polyhedron called? **faces**
2. What shape are the faces, other than base, of a pyramid? **triangles**
3. What shape are the faces of a prism? **polygons**
4. How many bases does a pyramid have? A prism? **one; two**

LESSON 10-6
Puzzles, Twisters & Teasers
Crossword

Across

2. Name of the flat surface of a polyhedron.
5. Three-dimensional object with flat surfaces.
7. When two faces of a three-dimensional figure share a side they form this.
9. Figure with one polygon shaped base; the other faces are triangles that come to a point.

Down

1. A prism is named for the shape of this.
3. Three-dimensional figure with a circular base and surface that comes to a point.
4. A figure with two congruent circular bases.
6. The point where three or more edges meet.
8. A polyhedron with two congruent parallel bases.

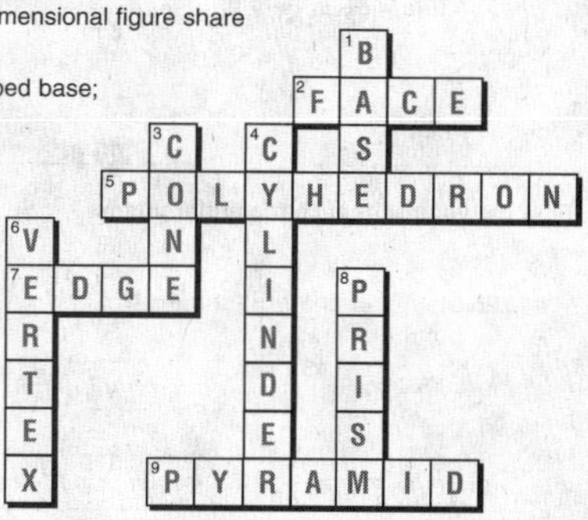

LESSON 10-7
Practice A
Volume of Prisms

Find the volume of each rectangular prism.

1. 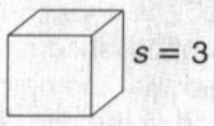

$V = 27$ in^3

2. 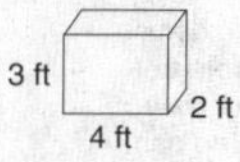

$V = 24$ ft^3

3. 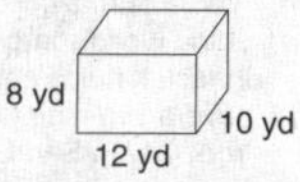

$V = 960$ yd^3

4. 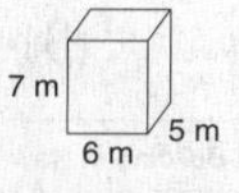

$V = 210$ m^3

5. 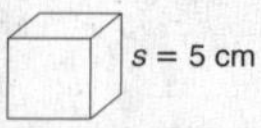

$V = 125$ cm^3

6. 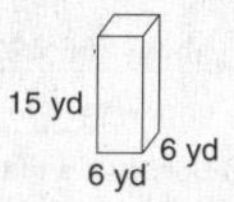

$V = 540$ yd^3.

Find the volume of each triangular prism.

7. 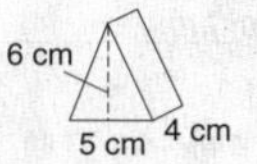

$V = 60$ cm^3

8. 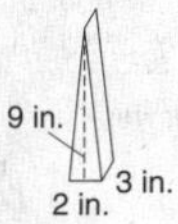

$V = 27$ in^3

9. 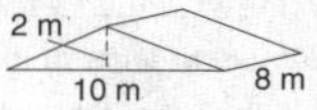

$V = 80$ m^3

10. A triangular prism and a rectangular prism have the same lengths, heights, and widths. Which prism has the greater volume?
the rectangular prism

11. Tim made a toy chest for his little sister's square building blocks. If 6 layers of blocks can fit in the box, and each layer has 15 blocks, how many building blocks can the toy chest hold in all?
90 building blocks

LESSON 10-7
Practice B
Volume of Prisms

Find the volume of each rectangular prism.

1. 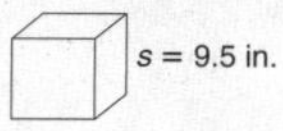

857.375 in^3

2. 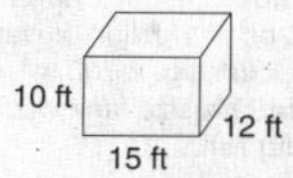

1,800 ft^3

3. 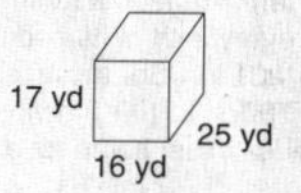

6,800 yd^3

4. 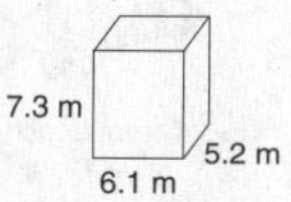

231.556 m^3

5. 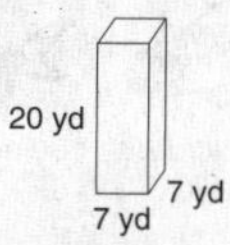

980 yd^3

6. 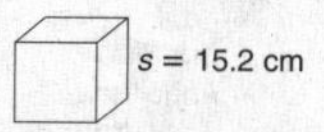

3,511.808 cm^3

Find the volume of each triangular prism.

7. 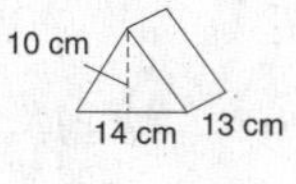

910 cm^3

8. 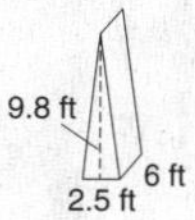

73.5 ft^3

9. 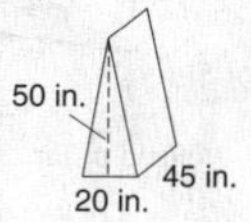

22,500 in^3

10. Fawn built a sandbox that is 6 feet long, 5 feet wide, and $\frac{1}{2}$ foot tall. How many cubic feet of sand does she need to fill the box?
15 ft^3

11. Unfinished lumber is sold in units called board feet. A board foot is the volume of lumber contained in a board 1 inch thick, 1 foot wide, and 1 foot long. How many cubic inches of wood are in 1 board foot?
144 cubic inches of wood

LESSON 10-7 Practice C
Volume of Prisms

Find the volume of each rectangular prism.

1.
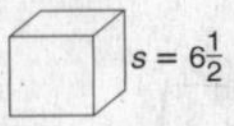

$274\frac{5}{8}$ in^3

2.
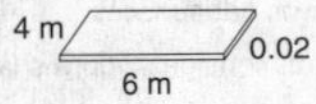

0.48 m^3

3.
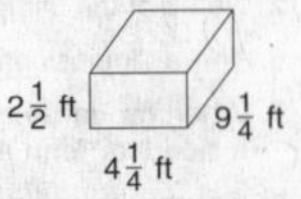

$98\frac{9}{32}$ ft^3

Find the volume of each triangular prism.

4.
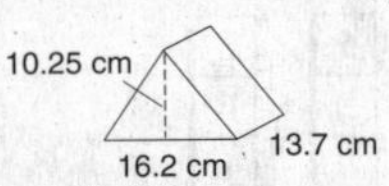

1,137.4425 cm^3

5.
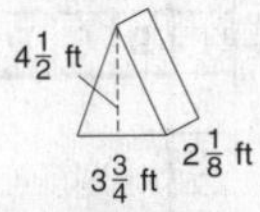

$17\frac{119}{128}$ ft^3

6.
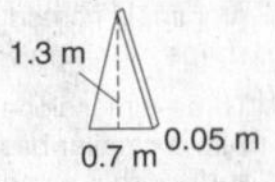

0.02275 m^3

Find the missing measurement for each rectangular prism.

7. ℓ = ? ; w = 34 m; h = 5 m; V = 340 m^3 **2 m**

8. ℓ = 4.5 ft; w = ? ; h = 6 ft; V = 256.5 ft^3 **9.5 ft**

9. ℓ = 14 yd; w = 8 yd; h = ? ; V = 1,344 yd^3 **12 yd**

10. ℓ = ? ; $w = 1\frac{1}{2}$ in.; $h = 3\frac{1}{2}$ in.; $V = 52\frac{1}{2}$ in^3 **10 in.**

11. A rectangular prism's base is 8 feet long. It is $2\frac{1}{2}$ times taller than it is long and $\frac{1}{2}$ as wide as it is tall. What is the volume of that prism? **1,600 ft^3**

12. A rectangular box is 3 feet wide, 6 feet long, and 4 feet tall. Inside the box, Phil placed a triangular prism that is 3 feet tall, and has a base 2 feet wide with a height of 5 feet. If he places another triangular prism of the same size in the box, how much empty space will there be in the box? **42 ft^3**

LESSON 10-7 Reteach
Volume of Prisms

Volume is the number of cubic units needed to fill a space. To find the volume of a rectangular prism, first find the area of the base.

length = 3 units
width = 2 units
$A = \ell w = 3 \cdot 2 = 6$ square units.

The area of the base tells you how many cubic units are in the first layer of the prism.

Next, multiply the result by the number of layers in the prism. The prism has 4 layers, so multiply 6 by 4.

$6 \cdot 4 = 24$

So, the volume of the rectangular prism is 24 cubic units.

Find the volume of each rectangular prism.

1.
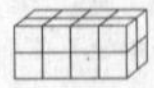

16 square units

2.
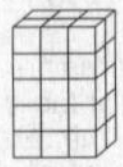

30 square units

To find the area of a triangular prism, first find the area of the base.

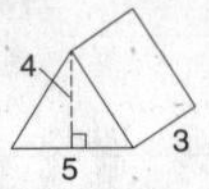

$A = \frac{1}{2}bh$
$= \frac{1}{2}(5 \cdot 4)$
$= 10$ square units

Then multiply the result by the height of the prism.

$10 \cdot 3 = 30$

The volume of the triangular prism is 30 cubic units.

Find the volume of each triangular prism.

3.
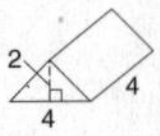

16 square units

4.
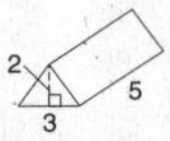

15 square units

LESSON 10-7 Challenge
Fish Tank Math

Did you know that most fish grow according to the size of their tank? In fact, a freshwater fish needs about 1 gallon of water for every inch of its body. The volume of 1 gallon of water is about 231 in^3. So, a fish tank with a volume of a 700 in^3 will hold $700 \div 231 \approx 3$ gallons of water. This size tank would be the healthiest home for a 3-inch-long fish.

Find the volume of each fish tank. Then match each fish to the best tank for it.

Angelfish ℓ = 6 in. | Koi Goldfish ℓ = 24 in. | Redtail Shark ℓ = 5 in. | Clown Loach ℓ = 12 in.

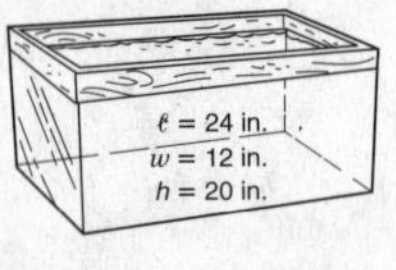

Volume of Tank: **5,760 in^3**
Gallons of Water: **about 25**
Fish: **koi goldfish**

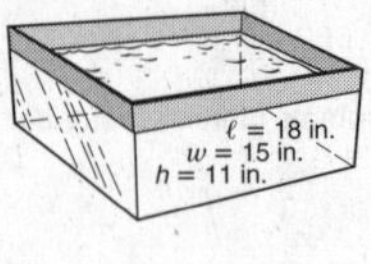

Volume of Tank: **2,970 in^3**
Gallons of Water: **about 13**
Fish: **clown loach**

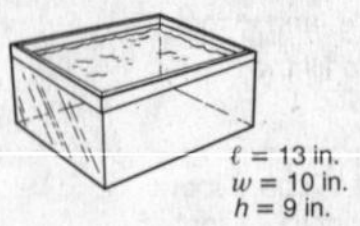

Volume of Tank: **1,170 in^3**
Gallons of Water: **about 5**
Fish: **redtail shark**

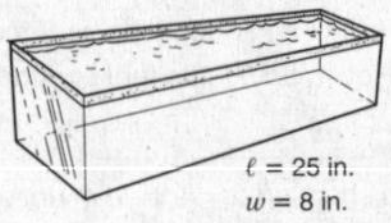

Volume of Tank: **1,600 in^3**
Gallons of Water: **about 7**
Fish: **angelfish**

LESSON 10-7 Problem Solving
Volume of Prisms

Write the correct answer.

1. At 726 feet tall, Hoover Dam is one of the world's largest concrete dams. In fact, it holds enough concrete to pave a two-lane highway from New York City to San Francisco! The dam is shaped like a rectangular prism with a base 1,224 feet long and 660 feet wide. About how much concrete forms Hoover Dam?
about 586,491,840 ft^3 of concrete

2. The Vietnam Veterans Memorial in Washington, D.C., is a 493.5-foot-long wall made of polished black granite engraved with the names of soldiers who died in the war. The wall is 0.25 feet thick and has an average height of 9 feet. About how many cubic feet of black granite was used in the Vietnam Veterans Memorial?
about 1,110.375 ft^3 of black granite

3. Benitoite, a triangular prism crystal, is the official state gem of California. One benitoite crystal found in California is 1.2 cm tall, with a base width of 2 cm and a base height of 1.3 cm. How many cubic centimeters of benitoite are in that crystal?
1.56 cm^3 of benitoite

4. The Flatiron Building in New York City is a triangular prism. A solid bronze souvenir model of the building is 5 inches tall, with a base height of 1.5 inches and a base width of 2.5 inches. How much bronze was used to make the model?
9.375 in^3 of bronze

Circle the letter of the correct answer.

5. Individual slices of pizza are sold in 2-inch-tall triangular prism boxes. The box base is 8 inches wide, with a 7-inch height. How many cubic inches of pizza will fit in each box?
A 112 in^3
B 102 in^3
C 60 in^3
(D) 56 in^3

6. The world's largest chocolate bar is a huge rectangular prism weighing more than a ton! The bar is 9 feet long, 4 feet tall, and 1 foot wide. How many cubic feet of chocolate does it have?
F 13 ft^3
G 14 ft^3
(H) 36 ft^3
J 72 ft^3

7. A box can hold 175 cubic inches of cereal. If the box is 7 inches long and 2.5 inches wide, how tall is it?
A 25 in.
(B) 10 in.
C 17.5 in.
D 9.5 in.

8. A triangular prism used to reflect light is made of 120 cm^3 of glass. If the prism is 5 centimeters tall, what is the area of each of its triangular bases?
F 24 cm
G 12 cm
H 12 cm^2
(J) 24 cm^2

LESSON 10-7 **Reading Strategies**
Analyze Information

The **volume** of a prism is the number of cubic units it contains.

Look at the first layer of cubic units in this rectangular prism.

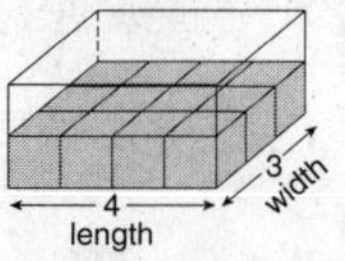

1. How many cubes long is the first layer? **4 cubes**
2. How many cubes wide is the first layer? **3 cubes**
3. How many cubes are in the first layer? **12 cubes**

Now look at the next layer of cubes.

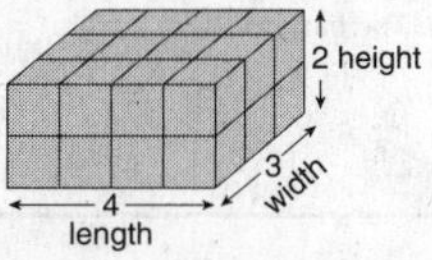

4. How many cubes long is the top layer? **4 cubes**
5. How many cubes wide is the top layer? **3 cubes**
6. How many cubes are in the top layer? **12 cubes**
7. What is the total number of cubes in both layers? **24 cubes**
8. What is the total volume of the rectangular prism? **24 cubic units**

LESSON 10-7 **Puzzles, Twisters & Teasers**
3D Match

Find the volume of each of the prisms below. Draw a line to the correct answer. Rearrange the letters to solve the riddle.

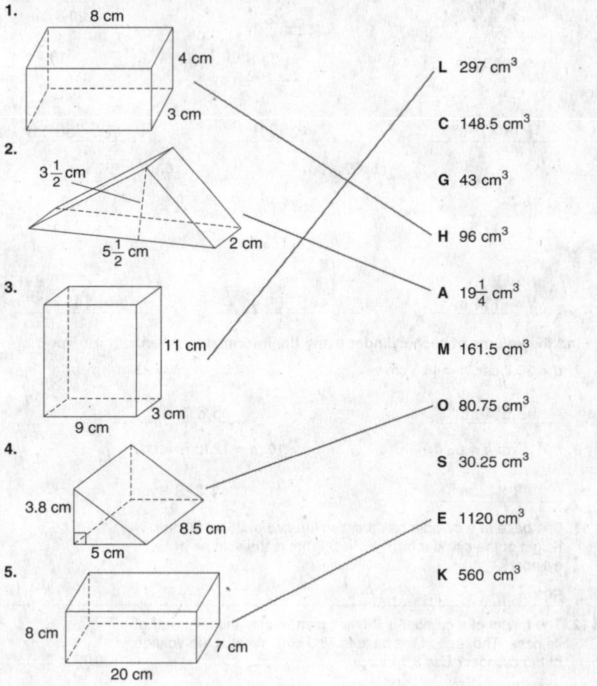

What grows the more you take away?

A hole

LESSON 10-8 **Practice A**
Volume of Cylinders

Find the volume *V* of each cylinder to the nearest cubic unit.

1. 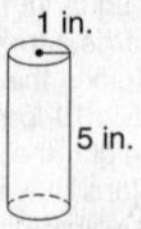

$V \approx 16$ in³

2. 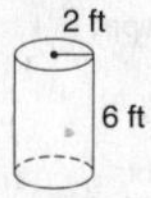

$V \approx 75$ ft³

3. 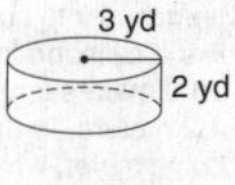

$V \approx 57$ yd³

4. 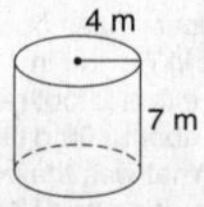

$V \approx 352$ m³

5. 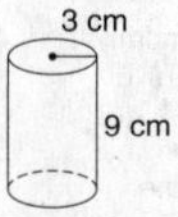

$V \approx 254$ cm³

6. 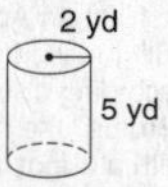

$V \approx 63$ yd³

7. 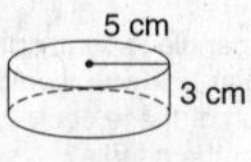

$V \approx 236$ cm³

8. 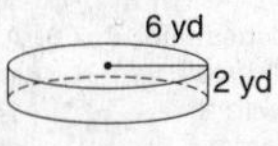

$V \approx 226$ yd³

9. 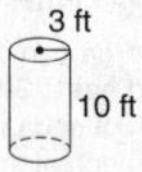

$V \approx 283$ ft³

10. Which of the following could be the measurement for the volume of a can: 20 centimeters, 20 cubic centimeters, or 20 square centimeters?

20 cubic centimeters

11. The diameter of a tube is 4 centimeters. It is 40 centimeters tall. What is the volume of the tube?

about 502.4 cm³

LESSON 10-8 **Practice B**
Volume of Cylinders

Find the volume *V* of each cylinder to the nearest cubic unit.

1. 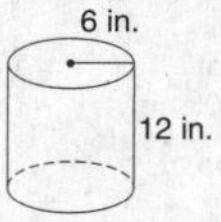

$V \approx 1{,}356$ in³

2. 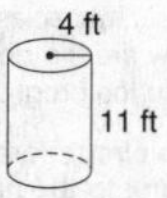

$V \approx 553$ ft³

3. 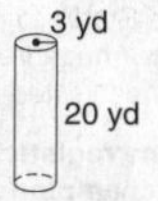

$V \approx 565$ yd³

4. 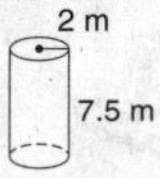

$V \approx 94$ m³

5. 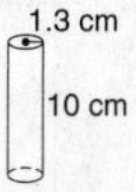

$V \approx 53$ cm³

6. 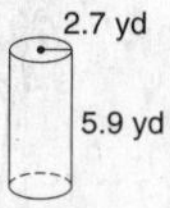

$V \approx 135$ yd³

7. 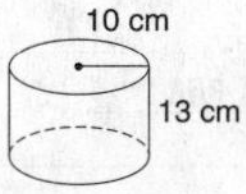

$V \approx 4{,}082$ cm³

8. 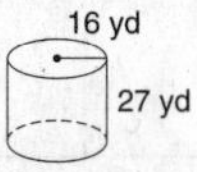

$V \approx 21{,}704$ yd³

9. 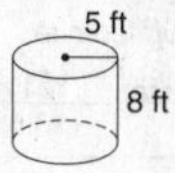

$V \approx 628$ ft³

10. A cylindrical package of oatmeal is 20 centimeters tall. The diameter of its base is 10 centimeters. About how much oatmeal does the package hold?

about 1,570 cubic centimeters of oatmeal

11. The volume of a can is about 50.24 in³. The radius of its base is 2 inches. How tall is the can?

4 inches

LESSON **10-8**

Practice C
Volume of Cylinders

Find the volume *V* of each cylinder to the nearest cubic unit.

1. 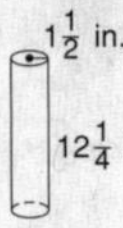

$V \approx 87$ in^3

2. 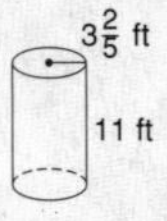

$V \approx 399$ ft^3

3. 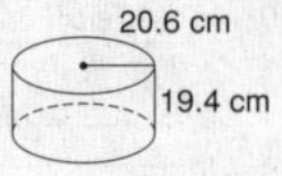

$V \approx 25{,}850$ cm^3

4. 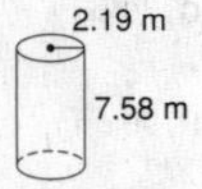

$V \approx 114$ m^3

5. 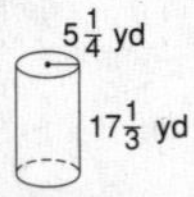

$V \approx 1{,}500$ yd^3

6. 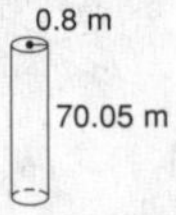

$V \approx 141$ m^3

Find the volume of each cylinder using the information given.

7. $d = 39.2$ cm; $h = 48.5$ cm

58,504 cm^3

8. $r = 8\frac{1}{2}$ in.; $h = 25$ in.

5,672 in^3

9. $r = 15$ m; $h = 68$ m

48,042 m^3

10. $d = 12$ ft; $h = 11.7$ ft

1,323 ft^3

11. The base of a cylinder has a circumference of 25.12 inches. The height of the cylinder is 10 inches. What is the volume of the cylinder?

502.4 in^3

12. The height of a cylinder is 3 times greater than the diameter of its base. The area of the base is 78.5 cm^2. What is the volume of the cylinder? Use 3.14 for π.

2,355 in^3

LESSON **10-8**

Reteach
Volume of Cylinders

You can use what you know about area to help you find the volume of a cylinder.

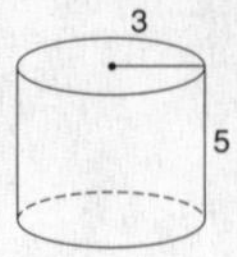

To find the volume of a cylinder, first find the area of the circular base.

$A = \pi r^2$ Use 3.14 for π.

$\approx 3.14 \cdot 3^2$

$\approx 3.14 \cdot 9$

≈ 28.26 square units

The area of the circular base is about 28.26 square units.

Next, multiply your answer by the height of the cylinder. The height of the cylinder is 5 units, so multiply 28.26 by 5.

$28.26 \cdot 5 = 141.3$

So, the volume of the cylinder is about 141.3 cubic units.

Find the volume of each cylinder.

1. 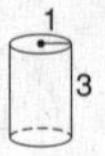

9.42 cubic units

2. 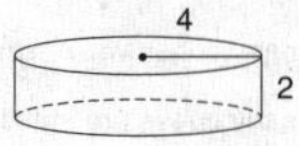

100.48 cubic units

3. 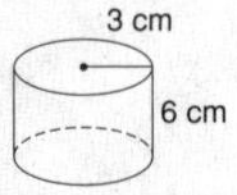

169.56 cubic units

4. 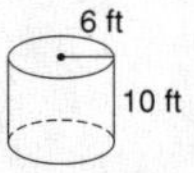

1,130.4 cubic units

LESSON **10-8**

Challenge
Champion Trees

Trees are among Earth's largest and oldest living things. The National Register of Big Trees measures trees throughout the United States to find the largest of each species—the champions. The four huge trees described below are the only champions that have been listed in the register since the program began in 1940.

Use the register's data to find the circumference and volume of each champion American tree trunk to the nearest foot and cubic foot.

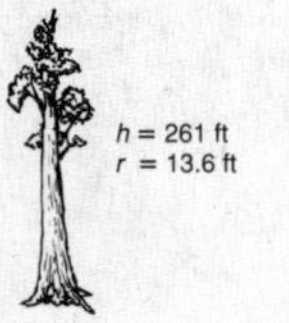

"General Sherman"
Giant Sequoia, California

Circumference: 85 ft

Volume: 151,582 ft^3

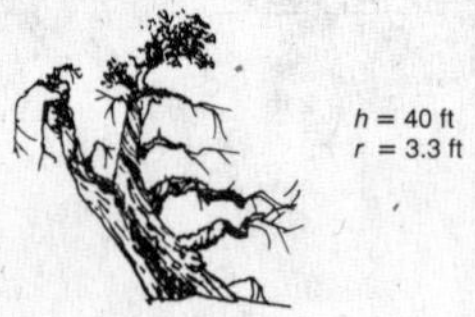

"Jardine Juniper"
Rocky Mountain Juniper, Utah

Circumference: 21 ft

Volume: 1,368 ft^3

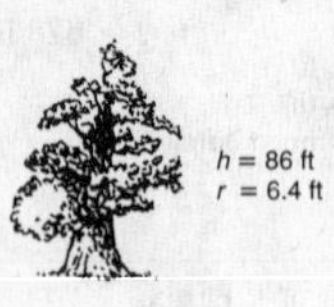

"Bennett Juniper"
Western Juniper, California

Circumference: 40 ft

Volume: 11,061 ft^3

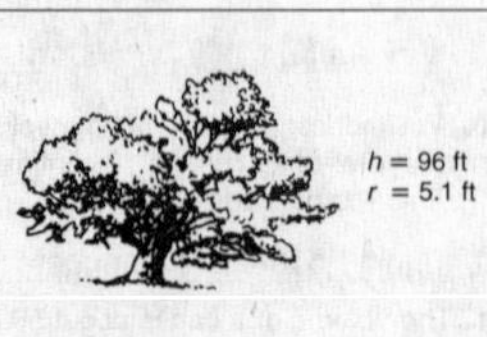

"Wye Oak"
White Oak, Maryland

Circumference: 32 ft

Volume: 7,840 ft^3

LESSON **10-8**

Problem Solving
Volume of Cylinders

Write the correct answer.

1. The Hubble Space Telescope was launched into space in 1990. Shaped like a cylinder, the telescope is 15.9 meters long, with a diameter of 4.2 meters. To the nearest whole cubic meter, what is the volume of the Hubble Space Telescope?

about 220 m^3

2. The Living Color aquarium in Bermuda is the largest freestanding cylindrical aquarium in the Western Hemisphere. With a 10-foot diameter and an 18-foot height, the aquarium holds 10,400 gallons of water! What is the aquarium's volume in cubic feet?

1,413 ft^3

3. In 1902 an American music company built the world's largest music recording cylinder. Nicknamed "Brutus," the cylinder is 5 feet tall, with a 2-foot diameter. What is the volume of the "Brutus" cylinder?

15.7 ft^3

4. The world's largest glass of orange juice was filled in Florida in 1998. At 8 feet tall and with a 2-foot radius, the glass held about 700 gallons of orange juice. What was the volume of that huge glass of orange juice?

100.48 ft^3

Circle the letter of the correct answer.

5. A large can of soda is 7.5 inches tall and has a 3-inch diameter. A small can of soda is 5 inches tall with a 2.5-inch diameter. To the nearest cubic inch, how much more soda does the large can hold?

A 53 in^3 more soda
(B) 28 in^3 more soda
C 25 in^3 more soda
D 20 in^3 more soda

6. A cylindrical candle is tightly packed in a rectangular box with a volume of 144 in^3. Which of these could be the dimensions of the candle?

F $h = 6$ in.; $r = 3$ in.
G $h = 2$ in.; $r = 5$ in.
(H) $h = 4$ in.; $r = 3$ in.
J $h = 3$ in.; $r = 4$ in.

7. The maximum length for an official professional baseball bat is 36 inches. Its maximum diameter is 2.6 inches. To the nearest cubic inch, what is the maximum volume of a professional baseball bat?

A 21 in^3
B 119 in^3
(C) 191 in^3
D 764 in^3

8. A can of tennis balls is 21 centimeters tall and has a diameter of 8 centimeters. What is the volume of the tennis ball can?

F 17,408.16 cm^3
(G) 1,055.04 cm^3
H 527.52 cm^3
J 263.76 cm^3

LESSON 10-8
Reading Strategies
Analyze Information

The volume of a cylinder is measured in cubic units.

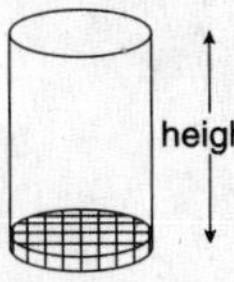

The area of the base gives the cubic units in the first layer. → *Area of circular base*

The height shows the number of layers. → *Area of base • height*

The base multiplied by the height is the volume of the cylinder. → *V = area of base • height*

1. What measurement shows the number of cubic units in the first layer of a cylinder?

   the area of its base

2. What measurement shows the number of layers in a cylinder?

   the height of the cylinder

3. How do you find the volume of a cylinder?

   Find the area of the base of the cylinder and multiply it by the height of the cylinder.

4. How do you find the area of the base of a cylinder?

   Use the formula for the area of a circle; $A = \pi r^2$

LESSON 10-8
Puzzles, Twisters & Teasers
Ice Ages

Find the volume of each cylinder. Match the letters to the correct answers to solve the riddle. Use 3.14 for π and round to the nearest tenth.

1. S 3,014.4

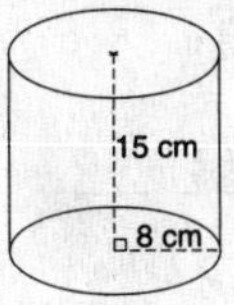

2. R 113

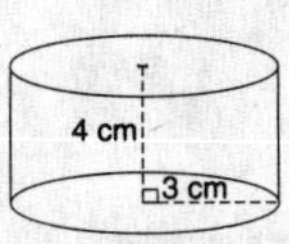

3. O 942

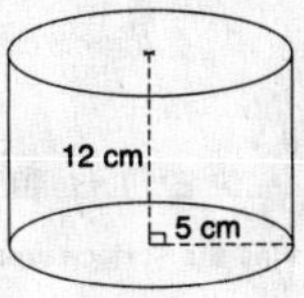

4. T 3,416.3

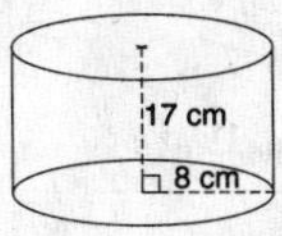

5. U 3,306.4

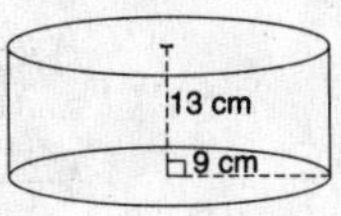

6. P 549.5

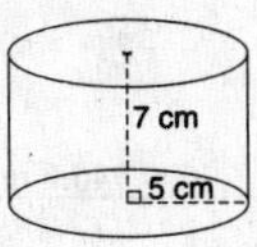

7. N 502.4

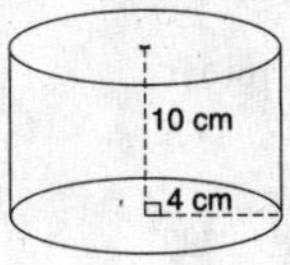

8. A 4,578.1

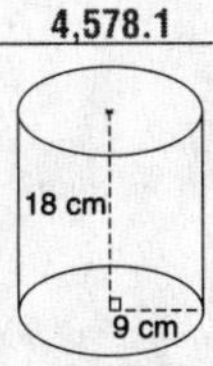

How does one dinosaur tell another to hurry up?

P	R	O	N	T	O
549.5	113	942	502.4	3,416.3	942

S	A	U	R	U	S
3,014.4	4,578.1	3,306.4	113	3,306.4	3,014.4

LESSON 10-9
Practice A
Surface Area

Find the surface area *S* of each net.

1.

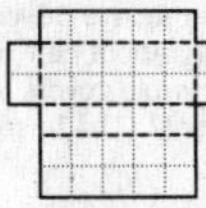

□ = 1 ft²

$S = 34\ ft^2$

2.

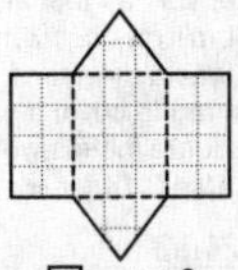

□ = 1 in²

$S = 34\ in^2$

3.

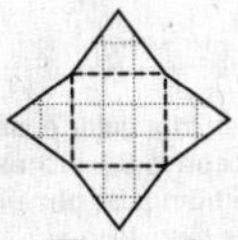

□ = 1 yd²

$S = 21\ yd^2$

4.

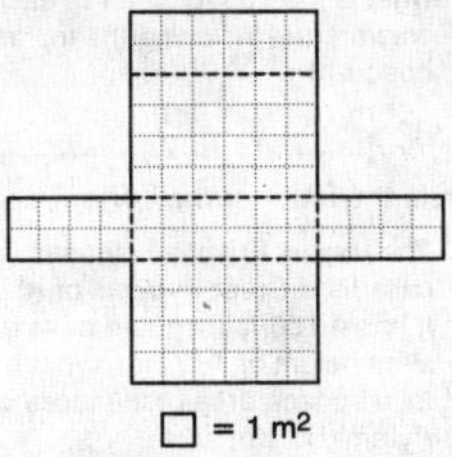

□ = 1 m²

$S = 88\ m^2$

Find the surface area *S* of each prism.

5.

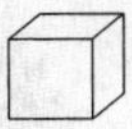

$S = 24\ in^2$

6.

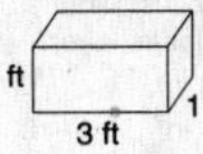

$S = 14\ ft^2$

LESSON 10-9
Practice B
Surface Area

Find the surface area *S* of each prism.

1.

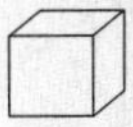

$S = 600\ in^2$

2.

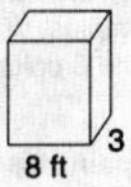

$S = 268\ ft^2$

Find the surface area *S* of each pyramid.

3.

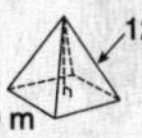

$S = 297\ m^2$

4.

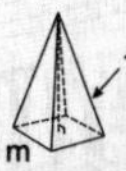

$S = 228\ m^2$

Find the surface area *S* of each cylinder. Use 3.14 for π.

5.

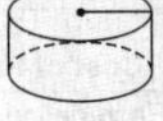

$S \approx 571.48\ cm^2$

6.

$S \approx 326.56\ in^2$

7. Why can you find an exact surface area measurement for a prism and pyramid but not for a cylinder?

   To find the area of a circle, you use *pi*, which is an estimated, not an exact value.

8. The surface area of a rectangular prism is 48 square feet. The area of its front is 4 square feet, and the area of one side is 10 square feet. What is the area of the top of the prism?

   $10\ ft^2$

LESSON 10-9
Practice C
Surface Area

Find the surface area *S* of each prism.

1.
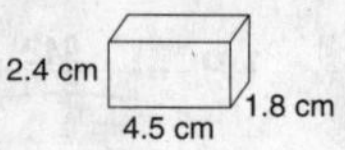

$S = 46.44 \text{ cm}^2$

2.
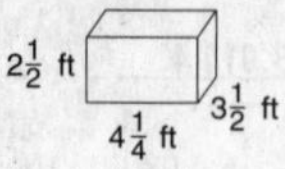

$S = 68.5 \text{ ft}^2$

Find the surface area *S* of each pyramid.

3.
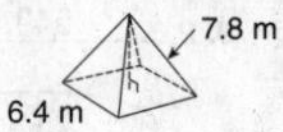

$S = 140.8 \text{ m}^2$

4.
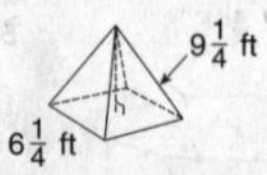

$S = 154\frac{11}{16} \text{ ft}^2$

Find the surface area *S* of each cylinder. Use 3.14 for π, and round to the nearest hundredth.

5.
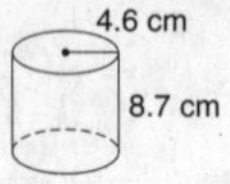

$S \approx 384.21 \text{ cm}^2$

6.
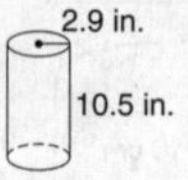

$S \approx 244.04 \text{ in}^2$

7. A cylindrical shipping box has a diameter of 4 inches and a height of 2 feet. What is the surface area of the box?

326.56 in^2 or 2.27 ft^2

8. A rectangular box has no top. It is 6 inches long, 4 inches wide, and 5 inches tall. What is the surface area of the box?

124 in^2

LESSON 10-9
Reteach
Surface Area

You can use what you know about finding the area of polygon to find the surface area of a solid figure.

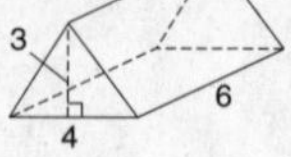

To find the surface area of the prism above, first find the area of each face.

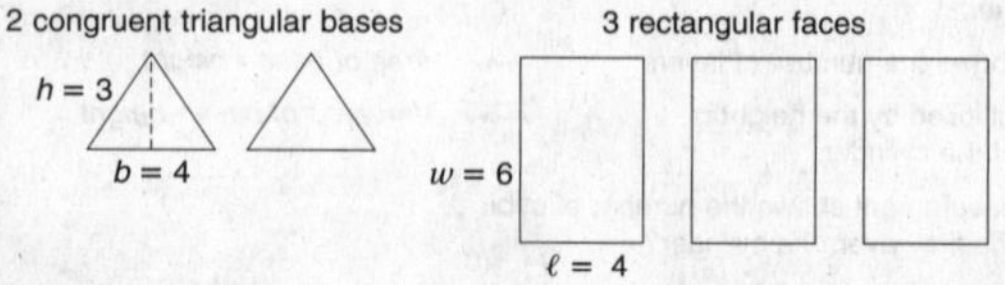

$A = \frac{1}{2}bh$
$= \frac{1}{2}(4 \cdot 3)$
$= 6$ square units

$A = \ell w$
$= 4 \cdot 6$
$= 24$ square units

Then, find the sum of all of the faces of the prism.

$SA = 6 + 6 + 24 + 24 + 24$
$= 84$ square units

Find the surface area of each figure.

1.
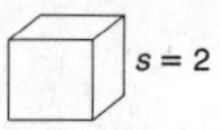

24 square units

2.
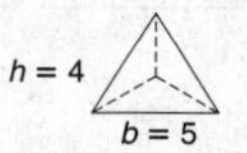

40 square units

LESSON 10-9
Challenge
A Monumental Paint Job

Completed in 1927, the Lincoln Memorial in Washington, D.C., honors President Abraham Lincoln. The monument was built in the style of Greek temples, with 36 columns surrounding the outside of the monument and 8 columns surrounding the statue of Lincoln inside.

It's your job to paint the columns of the Lincoln Memorial. You will paint its exterior columns blue, and its interior columns red. Use the diagram to figure out how much paint you should buy and how much the paint will cost.

each outside column:
$h = 44$ ft
$r = 3.7$ ft

each inside column:
$h = 50$ ft
$r = 3.75$ ft

Surface Areas to Paint	Gallons of Paint Needed	Cost of Paint
Remember to consider the parts of each column that will **not** be painted.	One gallon of paint will cover 350 square feet. Round **up** to the nearest whole gallon.	Each gallon of paint costs $16.99.
Exterior Columns: $36{,}805.824 \text{ ft}^2$	Exterior Columns: 106 gallons	Blue Paint: $1,800.94
Interior Columns: $9{,}420 \text{ ft}^2$	Interior Columns: 27 gallons	Red Paint: $458.73

LESSON 10-9
Problem Solving
Surface Area

Write the correct answer.

1. The world's largest cookie was baked in Wisconsin in 1992. Its diameter was 34 feet and contained about 4 million chocolate chips! If the cookie was a cylinder 1 foot tall, and you wanted to cover it with icing, how many square inches would you have to ice? Use 3.14 for π.

$276{,}721.92 \text{ in}^2$

2. The top of the Washington Monument is a square pyramid covered with white marble. Each triangular face is 58 feet tall and 34 feet wide. About how many square feet of marble covers the top of the monument? (The base is hollow.)

about $3{,}944 \text{ ft}^2$ of marble

3. The Parthenon, a famous temple in Greece, is surrounded by large stone columns. Each column is 10.4 meters tall and has a diameter of 1.9 meters. To the nearest whole square meter, what is the surface area of each column (not including the top and bottom)?

62 m^2

4. The tablet that the Statue of Liberty holds is 7.2 meters long, 4.1 meters wide, and 0.6 meters thick. The tablet is covered with thin copper sheeting. If the tablet was freestanding, how many square meters of copper covers the statue's tablet?

72.6 m^2 of copper

Circle the letter of the correct answer.

5. The largest Egyptian pyramid is called the Great Pyramid of Khufu. It has a 756-foot square base and a slant height of 481 feet. What is the total surface area of the faces of the Pyramid of Khufu?
   - (A) $727{,}272 \text{ ft}^2$ (circled)
   - B $727{,}722 \text{ ft}^2$
   - C $727{,}727 \text{ ft}^2$
   - D $772{,}272 \text{ ft}^2$

6. A glass triangular prism for a telescope is 5.5 inches tall. Each side of the triangular base is 4 inches long, with a 3-inch height. How much glass covers the surface of the prism?
   - F 6 in^2
   - G 12 in^2
   - H 39 in^2
   - (J) 78 in^2 (circled)

7. A can of frozen orange juice is 7.5 inches tall, and its base diameter is 3.5 inches. What size strip of paper is used for its label?
   - (A) 82.43 in^2 (circled)
   - B 26.25 in^2
   - C 576.98 in^2
   - D 101.66 in^2

8. Tara made fuzzy cubes to hang in her car. Each side of the 2 cubes is 4 inches long. How much fuzzy material did Tara use to make both cubes?
   - F 96 in^2
   - (G) 192 in^2 (circled)
   - H 16 in^2
   - J 128 in^2

LESSON **10-9**

Reading Strategies

Follow a Procedure

Surface area is the sum of the areas of all the faces of a solid figure.

If you unfold a solid figure and lay it flat, you make a net. A **net** is a pattern that makes it easy to see all faces of a solid figure.

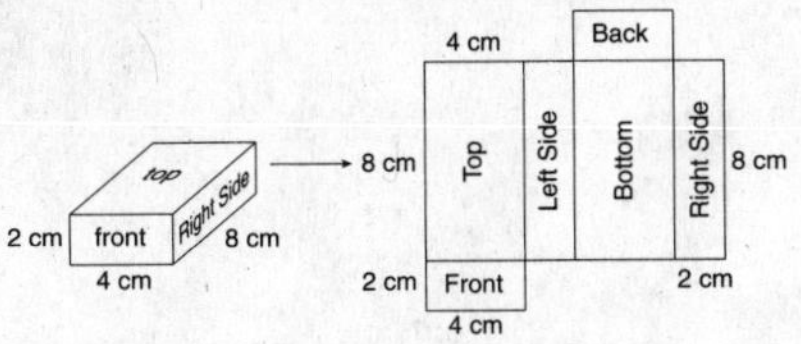

Follow these steps to find the surface area of this rectangular prism. Use the formula $A = \ell w$ to find the area of each face.

Step 1: Find the area of the top face. The bottom face has the same area.

↓

Step 2: Find the area of the right face. The left face has the same area.

↓

Step 3: Find the area of the front face. The back face has the same area.

↓

Step 4: Add the areas of all six faces to find the total surface area.

Answer the following questions.

1. What is the area of the top face? What face has the same area?

   $32\ cm^2$; the bottom face

2. What is the area of the right face? What face has the same area?

   $16\ cm^2$; the left face

3. What is the area of the front face? What face has the same area?

   $8\ cm^2$; the back face

4. What numbers do you add to find the total surface area of the prism? What is the total surface area?

   $32 + 32 + 16 + 16 + 8 + 8$; $112\ cm^2$

LESSON **10-9**

Puzzles, Twisters & Teasers

See Saw

For problems 1–4 use the triangular prism to answer each question. For problems 5–8 use the rectangular prism to answer each question. Choose the correct answer and write the letter in the space matching the problem number to solve the riddle.

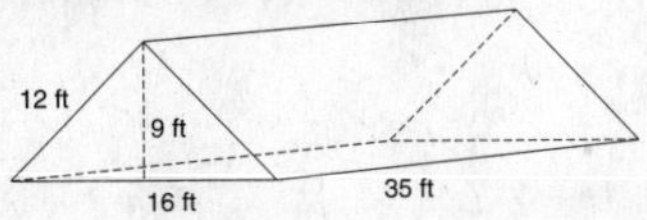

1. The area of the bottom of the triangular prism is...

   W = $37\ ft^2$ **R** = $72\ ft^2$ **L** = $144\ ft^2$ **(C)** = $560\ ft^2$ **A** = $315\ ft^2$ **D** = $420\ ft^2$

2. The area of the triangular side of the prism is...

   T = $37\ ft^2$ **(O)** = $72\ ft^2$ **E** = $144\ ft^2$ **L** = $560\ ft^2$ **B** = $315\ ft^2$ **S** = $420\ ft^2$

3. The area of the back face of the triangular prism is...

   D = $37\ ft^2$ **N** = $72\ ft^2$ **J** = $144\ ft^2$ **P** = $560\ ft^2$ **E** = $315\ ft^2$ **(A)** = $420\ ft^2$

4. The total surface area of the triangular prism is...

   M = $840\ ft^2$ **T** = $984\ ft^2$ **D** = $1,472\ ft^2$ **(I)** = $1,544\ ft^2$

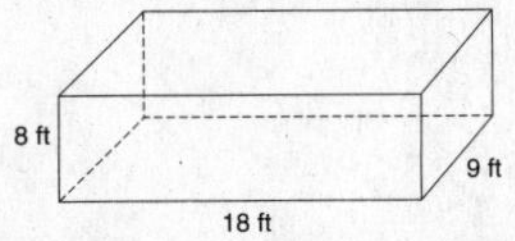

5. The area of the front face is...

   M = $52\ ft^2$ **A** = $72\ ft^2$ **S** = $162\ ft^2$ **(N)** = $144\ ft^2$ **T** = $172\ ft^2$ **L** = $1,296\ ft^2$

6. The area of the right-hand face is...

   C = $52\ ft^2$ **(E)** = $72\ ft^2$ **K** = $162\ ft^2$ **A** = $144\ ft^2$ **B** = $172\ ft^2$ **D** = $1,296\ ft^2$

7. The area of the bottom face is ...

   T = $52\ ft^2$ **A** = $72\ ft^2$ **(L)** = $162\ ft^2$ **O** = $144\ ft^2$ **G** = $172\ ft^2$ **C** = $1,296\ ft^2$

8. The total surface area of this rectangular prism is...

   (R) = $756\ ft^2$ **B** = $840\ ft^2$ **N** = $942\ ft^2$ **I** = $1,544\ ft^2$

What puts white lines on the ocean?

A	N		O	C	E	A	N		L	I	N	E	R
3	5		2	1	6	3	5		7	4	5	6	8